T0213998

SpringerBriefs in Computer Science

SpringerBriefs present concise summaries of cutting-edge research and practical applications across a wide spectrum of fields. Featuring compact volumes of 50 to 125 pages, the series covers a range of content from professional to academic.

Typical topics might include:

- A timely report of state-of-the art analytical techniques
- A bridge between new research results, as published in journal articles, and a contextual literature review
- A snapshot of a hot or emerging topic
- An in-depth case study or clinical example
- A presentation of core concepts that students must understand in order to make independent contributions

Briefs allow authors to present their ideas and readers to absorb them with minimal time investment. Briefs will be published as part of Springer's eBook collection, with millions of users worldwide. In addition, Briefs will be available for individual print and electronic purchase. Briefs are characterized by fast, global electronic dissemination, standard publishing contracts, easy-to-use manuscript preparation and formatting guidelines, and expedited production schedules. We aim for publication 8–12 weeks after acceptance. Both solicited and unsolicited manuscripts are considered for publication in this series.

**Indexing: This series is indexed in Scopus, Ei-Compendex, and zbMATH **

More information about this series at http://www.springer.com/series/10028

Bin Dong • Kesheng Wu • Suren Byna

User-Defined Tensor Data Analysis

 Springer

Bin Dong
Lawrence Berkeley National Laboratory
Berkeley, CA, USA

Kesheng Wu
Lawrence Berkeley National Laboratory
Berkeley, CA, USA

Suren Byna
Lawrence Berkeley National Laboratory
Berkeley, CA, USA

ISSN 2191-5768   ISSN 2191-5776   (electronic)
SpringerBriefs in Computer Science
ISBN 978-3-030-70749-1   ISBN 978-3-030-70750-7   (eBook)
https://doi.org/10.1007/978-3-030-70750-7

This Springer imprint is published by the registered company Springer Nature Switzerland AG
The registered company address is: Gewerbestrasse 11, 6330 Cham, Switzerland

Foreword

Bin Dong, Kesheng Wu, and Suren Byna have developed a highly readable and comprehensive book that provides the very first in-depth introduction to the interaction of two important topics of relevance to computational science: high-performance computing (HPC) and data analysis. The book provides background on both topics, but, more importantly for the first time in book form, also describes some of the more recent developments in data analysis using tensor formulation and the FT programming model that allows the efficient utilization of high-performance computing platforms. The book is unique by serving both as a reference book and a user guide to FT, as well as general introduction to big data systems, programming models, and data models.

It has been more than a decade since the publication in 2009 of the groundbreaking book "The Fourth Paradigm: Data-Intensive Scientific Discovery" [1] edited by Tony Hey, which celebrates in a number of essays the work of Jim Gray. Jim Gray's work as documented in this book set in motion an understanding of data-intensive computing as an essential part of the scientific discovery process, and thus has by now become an integral part of computational science. As a community, we have come to accept the notion that data-intensive computing is the tool to analyze the avalanche of data being produced by scientific experiments as well as numerical simulations. Some of the major scientific discoveries of the last decade, such as the Higgs boson at CERN, the incredible progress in exploring the genome, and understanding climate data as we face the challenge of climate change all have depended on data-intensive computing. The scientific computing community has worked diligently on refining algorithms, exploring new storage technologies, thinking about the challenge of distributed data analysis, and developing large software frameworks, so thus by now data-intensive computing has become a mature scientific activity.

In the same time frame, high-performance computing has seen even more dramatic developments. In the 2010s, computing technology has made a dramatic transition from MPPs using commodity hardware and the MPI programming model, to heterogeneous systems with GPU acceleration and an associated heterogeneous programming model, while increasing performance by a factor of one thousand from the Petaflops to the Exaflops level in 2021. Today, we are close to yet another transformation of the HPC field as the rapid advance of machine learning in the last decade has led to new hardware technology that addresses the computational kernels of machine learning algorithms directly–by relying on a tensor formulation.

In this context of this potential rapid transformation of the high-performance computing field, the book by Dong, Wu, and Byna arrives at exactly the right time. It succeeds perfectly and combines solidly the two almost parallel threads of developments in data-intensive computing and high-performance computing in one single volume. It will provide a solid foundation for anyone who is considering using a very recent tool such as FT in order to analyze complex simulation data or the ever-increasing amount of experimental data. I highly recommend this timely book for scientists and engineers. It closes an important gap by introducing a new tool for high-performance data analysis in the available literature on computational science. It will provide a solid reference as the community embarks on the Exascale data analysis adventure.

Lawrence Berkeley National Laboratory Horst Simon
Berkeley, CA, USA
May 2021

Preface

FasTensor is a generic parallel programming model for big data analyses with user-defined operations. FasTensor exploits the structural locality in the multidimensional arrays to automate file operations, data partitioning, communication, parallel execution, and common data management operations. This book introduces the FasTensor programming model and its C++ implementation. It also serves as the user guide to the software implementation with a detailed description of its application programming interface (API) along with illustrative examples.

This book starts with an introduction on the background and theoretical foundation for FasTensor. Many existing data management systems, including database systems and MapReduce systems, share the same data parallel and functional approach we use; however, most of these systems have made design choices that are not suitable for scientific applications. Due to this mismatch, some common scientific data analysis operations, such as convolution, could take thousand times longer than with FasTensor.

This book describes the C++ API from the current implementation of FasTensor. For many of the public functions, a sample code is provided to illustrate a common use case. The book also comes with two complete science examples from two different groups of domain scientists. In both cases, we are able to provide much more efficient tools for completing their analysis tasks with minimal programming efforts.

This book is intended to be a reference for FasTensor, so a reader could look up the most relevant piece of information when needed. However, it might be useful to read through the book to gain a broad overview of the context of the work and its design choices for scientific applications. Readers interested in applying FasTensor to their own use cases could jump to Chap. 3 or Chap. 4 to look into how to use FasTensor. This book uses notations adapted from C++ language and we assume the reader to have some familiarity with C++ programming language.

We integrated a significant amount of source code in the book to illustrate the usage of the FasTensor software. To get the most out of this book, we encourage the reader to try out these sample programs. Latest information about FasTensor could be obtained from https://sdm.lbl.gov/fastensor/

Berkeley, USA   Bin Dong

Berkeley, USA   Kesheng Wu

Berkeley, USA   Suren Byna

Acknowledgments

We would like to thank several contributors for their efforts to polish the idea of FasTensor. These contributors include Florin Rusu and Weijie Zhao from the University of California, Merced and Houjun Tang from Lawrence Berkeley National Laboratory (LBNL). We got lots of help from Quincey Koziol from LBNL about how to use HDF5 efficiently in FasTensor. We would like to thank Junmin from LBNL for her help in adapting FasTensor to work on ADIOS file formats. While we developed scientific use cases, we received lots of help from Patrick Kilian (previously at Los Alamos National Laboratory and now at The Space Science Institute), Gan Fuo (Los Alamos National Laboratory), Jonathan Ajo-Franklin (previously at LBNL and now at Rice University), Verónica Rodríguez Tribaldos (LBNL), and Xin Xing (Georgia Institute of Technology). We would like to thank our Scientific Data Management (SDM) group members, including Alex Sim, Vincent Dumont, Mariam Kiran, Bashir Mohammed, and Arie Shoshani, for their constant support and suggestions in improving the FasTensor R&D efforts.

We would like to extend our thanks to DOE program managers, Dr. Lucy Nowell and Dr. Laura Biven, for their consistent support of the R&D of FasTensor. We thank the IPO office at LBNL, who helped negotiating the contract of this book with Springer. This effort is supported by the U.S. Department of Energy (DOE), Office of Science, and Office of Advanced Scientific Computing Research under contract number DE-AC02-05CH11231. This work used resources of the National Energy Research Scientific Computing Center (NERSC), a DOE Office of Science User Facility that is supported under contract number DE-AC02-05CH11231.

The authors are very thankful to their family members. In specific, Bin Dong, the lead author, would like to thank his family members, Kainan and Yijun, for their patience, while he is sitting in front of computer working on this book. Kesheng Wu would like to Weihong, William, Elizabeth, and Winnie for their support during the preparation of this book. Suren Byna would like to thank Sujana, Pranav Shrey, and Pranga Suren, the family members supporting him in every aspect of his life.

Contents

Chapter 1
Introduction

The success of an online business depends on its ability to gather and process a large amount of data. Similarly, many of the headline grabbing scientific discoveries also rely on their ability to process vast quantities of data [2–4]. For example, the discovery of Higgs boson, also known as the god particle,[1] was based on a large physics experiment that produces data at the speed of 600 terabytes per second,[2] while the discoveries involving gravitational waves not only processed a large amount of data but also integrated information from a number of astronomical catalogs.[3] In one study of the gravitational wave from a neutron star merger,[4] the signal from the Laser Interferometer Gravitational-Wave Observatory (LIGO) was augmented with data from the Fermi Gamma-ray Space Telescope, Gamma-ray Burst Monitor (GBM), the INTErnational Gamma-Ray Astrophysics Laboratory (INTEGRAL), and so on [4]. Figure 1.1 illustrates how these data sources help to narrow the location of the particular neutron star merger [4]. As in this astronomy example, many scientific data sets are rapidly growing as shown in Fig. 1.2. These large scientific data sets lead to exciting scientific discoveries, but also present many challenges. One clear need is an advanced data analysis tool capable of handling these data sets. This book is about one such tool named FasTensor.[5] In this introductory chapter, we explore the core concepts behind the latest generation of

---

[1] Higgs boson discovery: https://www.nationalgeographic.com/news/2012/7/120704-god-particle-higgs-boson-new-cern-science/.

[2] That is about 600 million records per second with each record containing about one megabyte according to their online publication https://home.cern/science/computing/processing-what-record.

[3] Discoveries from gravitational waves: https://www.newscientist.com/article/2150906-the-5-biggest-discoveries-from-the-hunt-for-gravitational-waves/.

[4] For a news release on this discovery, see for example https://www.space.com/38471-gravitational-waves-neutron-star-crashes-discovery-explained.html.

[5] **FasTensor** was previously named **ArrayUDF** in our research publications [5–8].

© The Author(s), under exclusive license to Springer Nature Switzerland AG 2021
B. Dong et al., *User-Defined Tensor Data Analysis*, SpringerBriefs in Computer Science, https://doi.org/10.1007/978-3-030-70750-7_1

1

Fig. 1.1 A sketch showing how different data sources are used to locate the position of a neutron star merger on the celestial sphere, where HL indicates information from LIGO alone, HLV for information from LIGO and Virgo, GBM for Fermi Gamma-ray Burst Monitor, and IPN for information from Interplanetary Network including Fermi and INTEGRAL. The acronyms are from Kaslkiwal et al. [4], which has additional information about localization of the source of gravitational waves

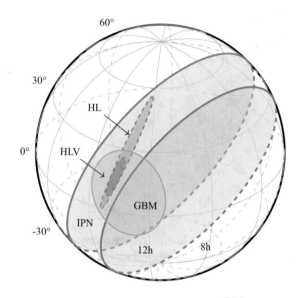

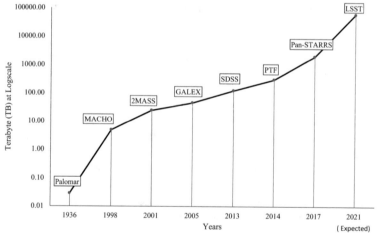

Fig. 1.2 Data size for a few well-known astronomy projects. Collectively, these projects are expected to collect over 100 exabyte in near future [9–14]

data processing systems. In later chapters, we will discuss the design trade-offs for developing a more science focused big data capability, and describe the application programming interface in detail.

1.1 Lessons from Big Data Systems

A key strategy for processing the large-scale data from above examples is to parallelize these tasks and harness the power of many computer processors. As recent as a decade ago, this type of parallel processing required dedicated custom computer hardware and software, or very expensive parallel database systems. However, along with the growth of internet businesses, a data processing revolution has emerged and the technology is quickly spreading [1, 15]. The core technology is the Big Data paradigm exemplified by the MapReduce (MR) system [16], which has inspired many variations and extensions, including Apache Hadoop [17], Apache Spark [18, 19], and TensorFlow [20]. The critical success factor of all these data processing systems is that they enable complex data analyses without requiring users to prescribe details of parallel execution, data management, or error handling. These data processing systems could be thought of as descended from database management systems (DBMS) that share this crucial objective, however, these DBMSs impose additional conditions making them far less attractive compared to the Big Data systems. For example, DBMSs require users to specify precise data schema and take full control of user data by creating their own copies [21, 22]. In addition, these parallel database management systems are quite expensive to purchase and operate. Scientific research projects and startup companies are often starved for funding and want to avoid these expensive data processing systems.

In large science projects, a common practice has been to develop their own custom data processing systems using open-source tools, such as ROOT [23, 24]. In addition, many tools have been built from lower level libraries including data communication libraries [25, 26]. Even though these data processing tools can be successful in a community, they often failed to impact other communities. For instance, ROOT is used by nearly all high-energy physics projects for the last two decades, however it has not been picked up by other scientific communities. In this case, one common complaint about the ROOT system is its steep learning curve.[6] There has been lengthy discussions on how to expand the impact of tools such as ROOT, but the proposed solutions are not widely accepted because these domain-specific tools are inherently hard to used in another domain. Following the successes of the Big Data systems, we propose to develop a high-performance data processing tool for common types of scientific data analyses.

At the conceptual level, a Big Data system has two interrelated components: a data model and a programming model. Next, we introduce the design choices of data model and programming model suitable for frequently used scientific analysis tasks. We see that the existing implementations of Big Data systems have chosen both data model and programming model that are not well-suited for scientific applications, which highlight the need for a new set of choices and a new software system.

---

[6]For a more thorough discussion on this topic, please refer to http://insectnation.org/articles/problems-with-root.html.

1.2 Data Model

In a Big Data system, the data model refers the logical data organization. For example, the byte stream, as codified in the POSIX I/O standard, is one of the oldest data model, widely used by file systems [27–29]. The byte stream data model allows I/O systems to read and write user data as a sequence of bytes. In this case, I/O does not care about the content of the bytes and an additional software layer is needed to manage the semantics of the data. To handle such semantic information in scientific applications, different community has adapted different high-level I/O libraries, such as FITS in astronomy [30], netCDF in climate research [31], and HDF5 in satellite imaging [32].

The Key-Value pair data model is a popular model used by many Big Data systems [15, 17, 19, 33], it is also at the core of the next generation file systems [34]. Another commonly used data model is the array data model. The high-level I/O libraries mentioned above, such as HDF5 and netCDF, follow this data model. In addition, popular data processing engines like TensorFlow [20], SciDB [35] and RasDaMan [36] also follow this data model. Large scientific data sets are often stored as arrays [1, 37], which explains why the high-level I/O libraries adapted by different scientific communities are all for storing arrays [31, 32, 38, 39].

The relational data model used by database management systems (DBMS) can be regarded as a version of this key-value pair data model with an additional layer of semantics about the data table. Even though a relational data table has rows and columns, one might think of it as a two-dimensional array. However, the relational algebra only offers options to access a row through its primary keys or through queries, users could not access a row or an element directly as in the array data model. This convenience in addressing an individual array element offers critical advantages. For example, TensorFlow (based on the array data model) can complete the same task faster than Apache Spark (based on the key-value pair data model) because the array data model allows TensorFlow to express key arithmetic operations more conveniently [7, 8]. Take the operation namely convolution[7] as example, it could be easily expressed in the array data model, but would require extensive searches and joins using the key-value pair data model, as illustrated by Fig. 1.3. Since scientific data analyses often involves complex operations such as convolution, it is critical to choose an appropriate data model [8, 40]. Tests shown that choosing the wrong data model could cause the operations to take thousands of times longer in execution time [7, 8]. Based on published literature, we have chosen to use the array data model for our science focused data processing system FasTensor. More details about the arrays in FasTensor will come in the following chapters of this book.

---

[7]The convolution operation is widely used to build the convolutional neural network (CNN), which demonstrates the efficacy of deep learning.

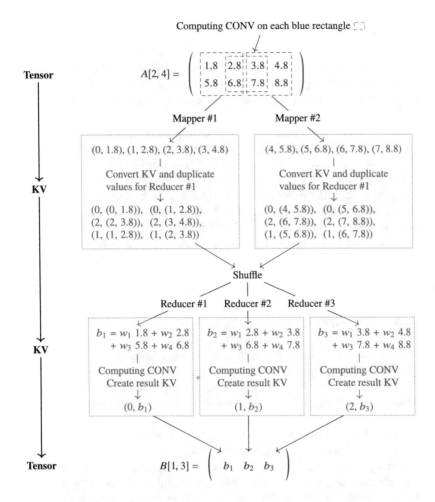

Fig. 1.3 An illustration of a 2 × 2 convolution on a 2 × 4 array A with a MapReduce system. The variables w_1, w_2, w_3 and w_4 are the weights for the convolution kernel. With the array notation, the convolution could be expressed as a series of multiplications and additions, while with the key-value (KV) pair data model, the eight elements of A have to be converted to eight key-value pairs and reorganized into three groups before the convolution computations (CONV). With a MapReduce system, considerable amount of time is needed to firstly linearize the array A into a 1D KV list and then create the three groups. This process not only involves multiple rounds of map and reduce operations, but also duplicates the values of A and introduces multiple keys where each key itself could take more space than a value of A

1.3 Programming Model

After deciding the data model, the next step in designing a Big Data system is to select a programming model. There are many different ways of constructing a parallel program [41–43]. The common approaches include: threading [41–44], communicating sequential processes [42, 43, 45, 46], functional programming [47–51], data parallel programming [49, 52], among others [53–55]. Most of these approaches need the programmer to decide how to lay out the arrays, and may even require the data communication to be explicitly managed by the user program. The exception is the functional programming approach that leaves the responsibility of data management to the run-time system. This approach is sometimes referred to as declarative programming [50, 50, 56]. It is codified as a fundamental principle of database system design [21, 22]. The Big Data systems generally follow this principle and offer a simple programming model where a user describes the operations on a single unit of data, such as, a key-value pair, or a collection of data items, such as a set of key-value pairs, without considering how the data records are loaded into memory, which processors are to perform the computation, which processors are to store the result of the computation, how to communicate among the processors, or how to recover from failures.

Under the same abstract data model and functional programming approach, the Big Data systems made drastic simplifications among the available choices. Many of the existing Big Data systems follow the original choices made by the MapReduce implementation [15, 33]: using the key-value pair as the basic data model and programming the analysis as a series of map and reduce functions. These choices allow the Big Data systems to significantly simplify their implementation compared to the previous generations of data processing systems. For example, parallel DBMS allows users to define their own data schema, while MapReduce system treats all user data as key-value pairs. This apparent primitive data model has enough information to allow users to define useful functions to implement any algorithms. A simpler data model, such as the byte stream, would make it much harder to construct a programming model without an additional semantic layer to provide meaning for the data. We believe array could replace the key-value pair as the data model of an effective Big Data system as demonstrated in TensorFlow.

The MapReduce programming model primarily allows users to specify two operations: the map function that transforms one key-value pair to another, and the reduce function that transforms a set of key-value pairs to a single key-value pair. This programming model is a much simplified version of the declarative programming (or functional programming) approach used in more traditional data processing systems. For this reason, it is relatively straightforward for database management systems to support for map and reduce operations [57, 58]. Because the functions supported by a Big Data system is so simple, it is straightforward for the system to figure out how to divide the data for parallel execution, determine the communication requirement, schedule the tasks, and recover from errors. In contrast, DBMS supports more variety of operations with potentially much more complex data dependency, parallel DBMS have been relatively expensive to program and to maintain.

In the spirit of adopting a simplified functional approach, the FasTensor system decides to support only one user-defined operation on nearest neighbors. Because the most important influence on a physical process comes from nearby points, computations from scientific applications often access only these nearest neighbors, and these accesses typically follow simple patterns called stencils [59–61]. This stencil based computation pattern is not only widely used in simulation of scientific processes, but also frequently used in data analyses [62]. For example, the aforementioned convolution operations is defined on a relatively small group of neighboring points that can be viewed as a stencil, see Fig. 1.3. These computational stencils are generally compact and this compactness allows us to design a mechanism for quickly expressing the stencil computation and a run-time system to efficiently execute the user-specified computation [7, 8]. We refer to this compactness as Structural Locality in later discussion.

In the last decade or so, the most exciting development in data analysis techniques is the emergence of machine learning, especially deep neural networks [63–65]. The success of Big Data systems can largely attribute to their support for machine learning. The computational kernels in many machine learning methods, especially those based on neural network, are matrix computations. Among the Big Data systems, TensorFlow [20] has the reputation of being more efficient than others due to its array data model and its optimized matrix computations for different computing platforms, especially on Graphical processing units (GPU).

Following the notion used by TensorFlow, we define a "tensor" as a multi-dimensional array along with transformation operations. Our FasTensor system takes the array data model and enables the users to define an arbitrary transformation by writing the computation on a stencil in a user-defined function. This approach allows the user to concentrate on what is to happen on a stencil, without considering details of partitioning of data onto different processors, communicating data among the processors, or recovering from errors. Tests on a variety of scientific applications demonstrate that FasTensor can easily and efficiently support a large variety of tensor transformations [5, 6].

1.4 High-Performance Data Analysis for Science

In addition to the above considerations on the data model and the programming model, an effective data analysis tool for scientific applications also needs to fit in with the larger scientific computing ecosystem. This includes requirements such as being efficient on high-performance computing (HPC) systems, coexisting with a variety of popular data management tools, following the common practice of data handling and so on [37].

In scientific applications, many of the data producing and consuming programs are constructed as communicating sequential processes using data communication libraries such as MPI [25, 26]. Popular data management tools such as HDF5 [32] and ADIOS [38, 39] are based on the same data communication substrate too. Therefore, we have chosen to use the same communication interface to allow FasTensor to easily fit into the HPC ecosystem.

We've mentioned that a number of scientific communities have adapted high-level I/O libraries such as HDF5 [32], netCDF [31] and ADIOS [38, 39] to store their data. These communities often have many petabytes of existing data files and an extensive set of custom analysis tools. It is critical to preserve the data files and not converting them into another format. Traditionally data processing based on database management systems (DBMS) converts user data into their own special formats, while most of the newer data processing systems follow the in situ processing approach that leaves the user data as is. By avoiding making another copy of the user data, this in situ approach does not need to reorganize user data before processing, it could often produce results faster than the traditional DBMS approach [66–68]. Thus, we believe that FasTensor needs to adopt this in situ approach when interact with user data files.

FasTensor only supports one type of user-defined functions, however, it is able to compose many versions of this function into complex analysis tasks in the same way a Big Data system allows map and reduce functions to be composed into a larger matrix computation routine or a machine learning application. One critical mechanism for supporting this composition capability is an effective caching mechanism for intermediate states [5]. This caching mechanism takes advantage of the large amount of main memory available on a typical high-performance computing system and can efficiently handle large and complex communication tasks required by different steps of a computation kernel. In a number of tests, our implementation of common computational kernels with FasTensor even could outperform the highly optimized TensorFlow versions [7, 8].

By design, FasTensor is optimized for high-performance computing systems used by scientific communities. Its programming model is well suited for common types of scientific computation. As we will show in later chapters, it executes many sophisticated data analysis operations very efficiently.

In the remainder of this book, we provide a more detailed description of the key components of FasTensor in Chap. 2, define the application programming interface in Chap. 3, and describe a couple of recent scientific application use cases in Chap. 4.

Chapter 2
FasTensor Programming Model

In the previous chapter, we have introduced the motivation for a big data analysis system and its essential components: data model and programming model. We have also clearly stated the reasons why FasTensor chooses the multi-dimensional array as its data model. In this chapter, we continue to provide more details on the FasTensor's programming model using the multi-dimensional array data model. The crucial constructs of a programming model for the data analysis include an abstract data type and a set of generic operators. The abstract data type allows users to define input and output data structures that their data analysis functions use. The abstract data type is defined on the top of the array data model. The set of generic operators should allow users to formulate a workflow with a wide range of data analysis functions. Through the abstract data type and these generic operators, a standard protocol between users and a data analysis system is established. On one hand, users can format their data with the abstract data type and express their data analysis with these generic operators. On the other hand, based on the abstract data type and these generic operators, the data analysis system can easily build its functions for generic data management functions, parallelization, and other tasks.

For instance, the *MapReduce* programming model uses the key-value (KV) pair as the abstract data type. The MapReduce provides two generic operators, i.e., *Map* and *Reduce*, to express data analysis functions. The *Map* and *Reduce* execute user customized procedures on a list of KV pair instances. As described in the previous chapter, the MapReduce programming model [16] created a revolution in the way of developing algorithms to analyze data in the KV pair model. *MapReduce* enables users to build complex data analysis algorithms without requiring tedious development of parallel execution, data management, and error recovery, among various other challenging tasks. Unfortunately, the *MapReduce* programming model lacks the support for multi-dimensional array data model, i.e., the array data model [8, 69]. As the array data model is prevalent in scientific domains, such as 2D sky survey images [70] and 3D plasma fields [71], the FasTensor programming model targets to revolutionize analysis of multi-dimensional array data. FasTensor operates directly on the array model and provides an execution engine to run composite user-defined custom analysis functions on massively parallel computing systems efficiently.

In the remainder of this chapter, we will introduce the FasTensor programming model and its essential data structures in supporting large-scale array data analysis. These data structures include *stencils*, *chunks*, and *overlap* (also known as *ghost zone* or *halo region* in literature). We will then present FasTensor's main operator, i.e., *Transform*. We will also describe the FasTensor execution engine and then present an evaluation of FasTensor to demonstrate its performance benefits over a state-of-the-art MapReduce framework, i.e., Apache Spark.

2.1 Array

A multi-dimensional array [72, 73] can have n dimensions, expressed as

$$Dim = Dim_0, Dim_1, \ldots, Dim_{n-1}.$$

In this book, we also express it as

$$Dim_0 \times Dim_1 \ldots \times Dim_{n-1}$$

or

$$Dim_0 \ by \ Dim_1 \ldots \ by \ Dim_{n-1}.$$

Obviously, the multi-dimensional array is significantly different from the KV-pair list, which is only a 1D data structure. The array can have a set of attributes, which are denoted as

$$Attr = Attr_0, Attr_1, \ldots, Attr_{m-1}.$$

Both n and m are positive integers here. Each dimension in an array is a continuous range of integer values. Each element in the array is called a cell. The indices for the array cell, i.e., $[i_0, i_1, \ldots, i_{n-1}]$, are denoted with square brackets. All cells have the same data type for an attribute and different attributes may have different data types. Based on these definitions, the array can be viewed as a function mapping from the dimension space to the attribute value:

$$Array : [Dim_0, Dim_1, \ldots, Dim_{n-1}] \longmapsto < Attr_0, Attr_1, \ldots, Attr_{m-1} >$$

Taking a 2D array, named A as an example, $A[i, j]$ represents the value of a cell at the index of $[i, j]$. If we further assume that the array A has two attributes, x and y, we use $A[i, j].x$ (with the dot . symbol) to refer to the value of the attribute x at the index $[i, j]$ and the $A[i, j].y$ refers to the value of the attribute y at the index $[i, j]$.

In an array, indices are always ignored to save access overhead and storage space. To allow accurate accesses of cell values, the cells are stored in well-defined layouts, such as the row-major order and the column-major orders. The row-major order is popular in scientific data formats, such as HDF5 [74] and NetCDF [75], and its last dimension is stored contiguously in storage devices. In the row-major order, offsets from the beginning to the location of a cell at $[i_0, i_1, \ldots, i_{n-1}]$ is expressed as $\sum_{k=0}^{n-1} \prod_{l=k+1}^{d-1} Dim_l i_k$. In the FasTensor programming model and its major components discussed in the following parts of the book, we assume that array data are stored with the row-major order. But, the FasTensor programming model can work on any array data layout.

2.2   Abstract Data Type: Stencil

An abstract data type of a general programming model is used to represent both the input data and the output data of its generic data operators. On the other hand, the abstract data type is also used to partition an input data into small units, each of which can be accessed, scheduled and processed independently or concurrently. In FasTensor, its abstract data type is called *Stencil*, which is conceptually a subset of a multi-dimensional array. The *Stencil* is denoted with the symbol S in the following parts of this chapter. In the FasTensor programming model, the *Stencil* data type has a **base cell** (or called as *center*) and one or many neighborhood cells that can be accessed using *relative offsets* from the base cell.

Using the *Stencil*, generic operators of the FasTensor programming model can involve a corresponding array cell in an input array (or multiple input arrays) and its surrounding neighborhood cells [76]. It is worth noting that *Stencil* in FasTensor is a data structure that is fundamentally different from the stencil computing pattern in numerical computing systems [77]. The stencil computing pattern in numerical computing mostly involves both data access and computing. The *Stencil* data structure in FasTensor only provides a way for users to access neighborhood cells. In FasTensor, the data analysis operation (computing) is defined by an end-user during the *Transform* operator, which is introduced in the following subsection.

In an array, the base element in a *Stencil* can be identified with its index in the array and the neighbors are denoted with their indexes calculated in relation to the base element. To generalize this concept, we use an **absolute index** of an array cell to represent the *base* cell and use **relative offsets** to represent the neighborhood cells. To formulate a *Stencil*, we use a two-element tuple

$$(b, \vec{r}),$$

where b is the absolute index of the base cell and $\vec{r}$ is a vector containing relative offsets of all neighborhood cells. For instance, in a 2D array A, we represent a

Stencil with nine points as:

$$\Big((1, 1), \big((0, 0), (0, 1), (0, 2), (1, 0), (1, 1), (1, 2), (2, 0), (2, 1), (2, 2)\big)\Big)$$

In this equation, *(1,1)* is the absolute index of the base cell (i.e., the *b* in the tuple). The relative offsets ($\vec{r}$) represent nine neighborhood cells, which include the base cell itself and other eight cells at the right-bottom of the base cell. This *Stencil* is illustrated in Fig. 2.1c. Instead of this index based form for a *Stencil*, a more user-friendly expression is a value-based representation. Taking the same example as above, its value based form is shown below:

$$\Big(A[1, 1], \big(S_{0,0}, S_{0,1}, S_{0,2}, S_{1,0}, S_{1,1}, S_{1,2}, S_{2,0}, S_{2,1}, S_{2,2}\big)\Big)$$

In this equation, *A[1, 1]* represents the base cell at the *absolute index [1,1]* of the array A and the $S_{i,j}$ represents a neighborhood cell at the *relative offset (i,j)*. We show a few examples of Stencils in Fig. 2.1.

Detailed description for the implementation of the Stencil in FasTensor is presented in Sect. 3.2 of the following chapter. Here we summarize useful properties of the Stencil in the FasTensor programming model.

- The *Stencil* abstraction allows partitioning of a large array into subsets logically without losing the array's structural locality property. This logical partitioning here means that FasTensor does not use *Stencil* to partition an array into small units, but it only executes the generic operator on each of these units. In comparison to array databases [78] that use large chunks with fixed size, the Stencil in FasTensor programming model allows a flexible partitioning of

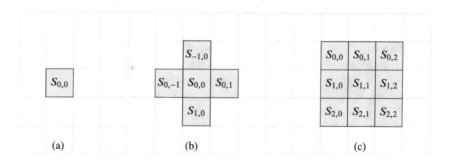

(a)   (b)   (c)

Fig. 2.1 A few *Stencil* examples in a 2D array. The rectangle filled with light red denotes the base cell. The rectangle filled with light blue denotes the neighborhood cells. $S_{i,j}$ represents the neighbor cells of the Stencil S at the relative offset (i, j). From the left to the right: (a) a single cell Stencil, which contains the only base cell without any neighbor; (b) a Stencil containing a center cell and four surrounding cells (left, right, above, and below). (c) a Stencil containing a center cell and its two-layers of neighbors only to the right, below, and to the corners. This type of Stencil can be used in computing convolutions in machine learning applications. Other popular Stencil examples and their applications can be found in our previous work [7]

neighborhood of corresponding array elements. Stencils also provide abstractions for using relative locality of array elements to present an array's structure. Although the FasTensor library also uses the chunk idea (as stated in the following parts of this chapter), the *Stencil* abstraction is independent from the chunk. The chunks in FasTensor are used to partition the array into small units to read input data, to schedule execution of an analysis operation, and to write output data. A *Stencil* dynamically builds input and output for a data analysis operation, which can be executed on a chunk or a whole array.

- Array cells in a Stencil may form any spatial shape in FasTensor. Using FasTensor, one can specify any base element and define the neighborhood using any valid relative offset. As shown in Fig. 2.1, neighborhood can be expressed in the middle or to the left or right most corners, etc. Therefore, the expressions of the Stencil in FasTensor are much more versatile than Spark [18, 79] or SciDB [73] or various SQL databases [80–84].
- The structure of a Stencil is flexible to be applied to the entire set of array elements. A Stencil's base element and it's neighborhood can be expressed with a simple representation relative to the base element. This flexibility allows keeping any number of desired cells for performing analysis.
- Multiple Stencils in FasTensor can share overlapping array element cells. As said above, the *Stencil* only perform logically partition without physically splitting the data for each Stencil. For instance, in a *4 by 4* 2D array A, let's define two Stencils $S1$ and $S2$ as below:

$$S1 = \left(A[0, 0], \left(S_{0,0}, S_{0,1}, S_{1,0}, S_{1,1} \right) \right)$$

and

$$S2 = \left(A[0, 1], \left(S_{0,0}, S_{0,1}, S_{1,0}, S_{0,1} \right) \right)$$

where the $S1$ has the base cell at $[0, 0]$ and the $S1$ has the base cell at $[0, 1]$. Both $S1$ and $S2$ have four neighborhood cells at the right-bottom of the base cell. Actually, $S1$ and $S2$ share the two cells: $A[0,1]$ and $A[1,1]$, which is shown in Fig. 2.2. In several scientific data analysis tasks, such as moving average calculation, overlapping data access to the same cells is a common pattern when using two consecutive cells. A stencil provides a logical partition without duplicating the data and therefore can efficiently support them executing on $S1$ and $S2$ concurrently.

2.3 Operator: Transform

The FasTensor programming model only has one operator namely *Transform*. The *Transform* operator executes a given user-defined function (UDF) on the Stencil abstraction type, which is defined in the previous section. Therefore, the FasTensor

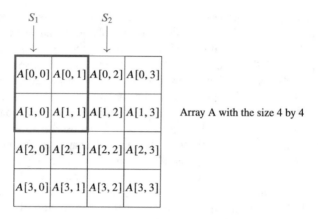

Fig. 2.2 Two Stencils S_1 (in red rectangle) and S_2 (in blue rectangle) defined on a 2D array A, which has the size 4 by 4. S_1 and S_2 share two array cells, $A[0, 1]$ and $A[1, 1]$, without any penalty

programming model can be represented as the following two-element tuple: *(S, Transform)*, where S is the *Stencil* data type and *Transform* is the generic operator. Given two arrays A and B,

$$A : [Dim_0, Dim_1, \ldots, Dim_{n_1}] \mapsto\, < Attr_0, Attr_1, \ldots, Attr_{m_1} >$$

$$B : [Dim_0, Dim_1, \ldots, Dim_{n_2}] \mapsto\, < Attr_0, Attr_1, \ldots, Attr_{m_2} >$$

where n_1, n_2, m_1 and m_2 are all positive integer, the generic operator *Transform* is expressed as,

$$Transform : \vec{S} \xrightarrow{\;f\;} \vec{S'}, \quad S \subset A, S' \subset B, \tag{2.1}$$

Semantically, *Transform* maps the *Stencil* abstraction $\vec{S}$ from array A to the result *Stencil* abstraction $\vec{S'}$ from the array B. The $\vec{S}$ represent neighborhood cells from the input array A and $\vec{S'}$ represents neighborhood cells from the output array B. The user customized function f defines the desired operation for the map. The $\rightarrow$ (hat) symbol on *Stencil* means both input and output can be a vector *Stencil* across all attributes of input and output arrays. In the following parts of this chapter, we uses $B = Transform(A, f)$ to represent the execution of the function f from A to B. Users can use the *Stencil* abstraction and its member cells to describe the desired operation in the function f. Users also control the return value of function f which is used to initialize the output *Stencil*.

A large array can be split into lots of subsets which are uniformly represented with the *Stencil* abstraction. Hence, one can easily create lots of *Stencil* instances which can be processed currently. This is similar to the MapReduce, where different KV pair lists can be processed by different mappers or reducers at the same time.

The FasTensor library in the following Chap. 3 provides an implement of such mechanism. Conceptually, the FasTensor library can automatically create multiple *Stencil* instances from an array and feed them into the user-defined function. The output Stencil instances from the user-defined function can also be automatically stored into proper places in the result array.

As stated in the previous subsection, the *Stencil* defines a subset of a multi-dimensional array. The *Transform* operator discussed in this section basically maps the subset of a multi-dimensional array to the subset of another multi-dimensional array. Therefore, the whole data programming model actually transforms data from an array to another. Such descriptions naturally match one definition of the tensor, which says that a tensor can be expressed with a multi-dimensional array with specific transformation rules. Hence, we call this new data programming model as FasTensor. In the following sections of this chapter, we present other data structures, including *chunk*, *overlap*, and the *execution engine* in FasTensor to support its applications in processing large arrays of data.

2.4 FasTensor Execution Engine

This section focuses on the parallel execution engine for the FasTensor programming model. We also introduce other components, such as chunking and overlap, of the execution engine. In Fig. 2.3, an high-level example of FasTensor's parallel execution is shown to process a 2D input array namely A. This example has two sets of *Transform* operations forming a simple workflow. The FasTensor execution engine uses the single program and multiple data (SPMD) paradigm [85], where multiple FasTensor processes are launched with the same analysis program and each process handles different partitions of the data. In Fig. 2.3, there are two processes running. These processes may be allocated on multiple cores of a single compute node or on multiple compute nodes. Within each process, a *Transform* operation executes a user-defined function. In this example, these two user-defined functions are denoted with f_1 and f_2.

In this example, each process executes a *Transform* operation with the f_1 function first. The initial data comes from an Array A, which is stored in a file in a parallel file system. The file in FasTensor is called endpoint, which has the format like HDF5, ADIOS and PnetCDF. The input array A is partitioned by FasTensor into multiple chunks. During this partitioning phase, FasTensor augments each chunk with overlap layers (also known as 'ghost zones' or "halo") to avoid communication required to access data from chunks that get partitioned to other processes. See more details about chunk and overlap in the following subsections. The output array of *Transform* with the f_1 function can either be cached in the distributed memory of compute nodes or be written (or called backup) to a persistent storage layer. Using the distributed memory to cache output of a *Transform* operation allows accessing the data fast by another *Transform* operation with the f_2 function. For two consecutive *Transform* operations, the latter operation may also access the overlap

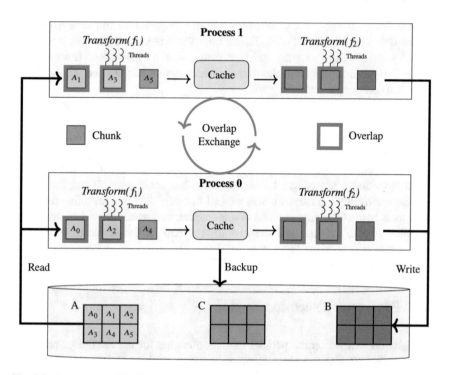

Fig. 2.3 An example of FasTensor execution of two user defined functions (f_1 and f_2) on a 2D array A through the *Transform* operator. Both f_1 and f_2 represent data analysis operations provided by users. The array A is partitioned into 6 (2×3) chunks and read by *two* processes, denoted as Process 0 and Process 1, for parallel processing. We used a simple round-robin scheduling algorithm here to allocate all 6 chunks of A among two processes. Specifically, each process works on three chunks, i.e., A_0 A_2 and A_4 on Process 0 and A_1, A_3 and A_5 on Process 1. These *two* processes can be created by MPI or other similar libraries on a single computing node or two computing nodes. Within each process, multiple threads can be launched to process a chunk of data concurrently. When array A is read, an overlap layer is appended to the chunks to avoid communication during the execution of the *Transform* operation. An intermediate array (C) from f_1 is cached locally in the main memory for f_2 to read. Before the actual read operation by f_2, an overlap exchange algorithm is executed to synchronize the overlap layer for all cached chunks. The intermediate array (C) can be backed up to disk for tolerating any failures. The final output array is B. All A, B and C are stored in a file system using self-describing file formats such as HDF5, netCDF, or ADIOS

layer. Thus, FasTensor's execution engine uses an overlap exchange algorithm that allows synchronizing each partitioned chunk. While storing an intermediate array in memory is prone to software or hardware failures, FasTensor can backup the intermediate array to a persistent storage layer, such as a parallel file system.

In most cases, the output array from a *Transform* operation has the same dimensions and the size as the input array. However, the output array may have different number of dimensions, size, and type from the input array. FasTensor automatically detects these parameters for an output array based on the information extracted from input the array and the user-defined function applied on it. The output from the final *Transform* operator is stored using efficient parallel writing operations. FasTensor currently supports writing the final output array into self-describing file formats, such as HDF5, PnetCDF, etc. A user can specify the file format using an endpoint definition to write an output array.

2.4.1 Chunk

The FasTensor library can process very large multi-dimensional arrays using parallel processing. In the case where a partition of array may not fit in the memory of a compute node, FasTensor allows splitting these large arrays into small **chunks** for parallel processing on multiple nodes or sequential processing on a single node A chunk can be defined using a set of *(starting indices, ending indices)*, where *indices* are vectors. Users can provide the chunk size to FasTensor, which is used for calculating the *(start indices, end indices)* set for all dimensions of an array. For instance, if chunk size is specified as $(2, 2)$ for a 2-D array of size $(4, 4)$, the first chunk contains cells between $(0, 0)$ and $(1, 1)$. After partitioning the input array's data into chunks, FasTensor's scheduling method assigns chunks among multiple parallel processes with the round-robin method.

Chunks from a large array are linearized using the row-major order and the linearized index is used as the 'ID' of the chunk denoted with id_{chunk}. Given there are p processes, the chunk with ID id_{chunk} is assigned to the process at rank $\lfloor id_{chunk}/p \rfloor$. An example of this scheduling is shown in Fig. 2.3, where the first chunk with the ID $A0$ is assigned to the first process and the second chunk with the ID $A1$ to the second process. Because of the SPMD paradigm, different processes can compute on their own chunk IDs and fetch them. By default, FasTensor assigns chunks of an array according to its linearized order. FasTensor also allows users to choose the assignment in the reverse direction, i.e., from the end to the beginning along linearized order. Within a single chunk, cells are also scheduled by their row-major order. More details of defining chunks, setting the chunk size, etc. are in Sect. 3.3.2 of the following chapter.

2.4.2 Overlap

In the execution of the FasTensor programming model on parallel computing environments, overlap regions (also known as "halo" or "ghost zones") are used to eliminates the need to retrieve cells from other chunks. It is a common practice to exchange overlap regions at synchronization points [86]. FasTensor allows augmenting a chunk with an overlap region, which contains cells surrounding a chunk in all dimensions. Users can specify the depth of the overlap region with the maximum offset used by the *Stencil* abstraction in the user-defined function f. Note that the overlap region in FasTensor introduces a small number of duplicated cells at the chunk boundary. These overlap regions are typically a small fraction of the chunk dimensions in most applications. While there are duplicated data operations to cover the overlap regions, they are much smaller in size than the data chunks and this redundancy has a negligible performance impact in FasTensor. More details of setting the overlap (or ghost zone or halo) are in Sect. 3.3.3.

2.5   Performance of FasTensor in Scientific Applications

To demonstrate performance benefits of using FasTensor over existing the MapReduce programming model, we use analysis tasks on data from two science areas: Community Earth System Model (CAM5) data [87, 88] and interaction of solar weather with magnetosphere simulation (VPIC) [40, 71]. We compare the performance of these analysis tasks using FasTensor and Apache Spark. The Apache Spark is the latest implementation of the MapReduce programming model with an advanced in-memory storage layer. Although Apache Spark is not originally developed for HPC systems, we used a couple of new technologies (i.e., H5Spark [89] and file pooling [79]) that improved Spark's usage on HPC systems.

We conducted a performance evaluation on the Cori supercomputer system located at the National Energy Research Scientific Computing Center (NERSC). Cori is a Cray XC40 system that has a peak performance of about 30 petaflops and debuted in 2017 as the world's fifth most powerful supercomputer in the world. We used the "Haswell partition" of Cori, where each node has a Intel Xeon "Hasewell" processor. The data was stored on a Lustre parallel file system and used the MPI programming model for running the tests. For Spark, data read operations are completed before the *Map* and *Reduce* operations start. To have a fair comparison between FasTensor and Apache Spark, we consider the computation time and execution time of these data analysis tasks separately because Spark is not originally designed and optimized for reading data from Lustre parallel file system. But the execution time can clearly show the performance benefits of FasTensor over Apache Spark.

2.5.1   Convolution Computations on a Climate Dataset

CAM5 is a widely used climate community model to study the global climate change [87]. The dataset we used here is generated by a 25 km spatial resolution climate simulation that simulates the years from 1979 to 2005. The 25 km resolution creates an image per a time step for the global atmospheric state with a 768×1152 matrix. The computation we used for comparing is convolutional neural network (CNN) [88], which was previously used to predict extreme weather events, such as atmospheric rivers (AR), from this data. Specifically, we have used FasTensor and Apache Spark to compose three key steps of CNN: *CONV*, *ReLU*, and *Pooling*.

```
//Three user defined functions
// take an input Stencil object iStencil and
// return an output Stencil object oStencil.
//The iStencil_{i,j,...} accesses the neighborhood cell at the offset (i, j, ...)
CONV_UDF( iStencil )
{
```

$$oStencil = \begin{bmatrix} iStencil_{0,0}, iStencil_{0,1}, \ldots, iStencil_{2,2} \end{bmatrix} \times \begin{bmatrix} w_{0,0} & \cdots & w_{0,k} \\ w_{1,0} & \cdots & w_{1,k} \\ \vdots & \vdots & \vdots \\ w_{8,0} & \cdots & w_{8,k} \end{bmatrix};$$

```
        return oStencil;
}

POOL_UDF( iStencil )
{
```
$$oStencil = max\left(\begin{bmatrix} iStencil_{0,0,0}, iStencil_{0,0,1}, \ldots iStencil_{0,0,2} \end{bmatrix}\right);$$
```
        return oStencil;
}


ReLU_UDF( iStencil )
{
```
$$oStencil = max\left(0, iStencil_{0,0,0}\right);$$
```
        return oStencil;
}

main()
{
        //Execution by chaining three Transforms on array A
```
$$R = Transform\left(Transform\left(Transform(A, CONV\_UDF), ReLU\_UDF \right), POOL\_UDF \right);$$
```
}
```

Fig. 2.4 Sample code of essential steps in CNN on a 2D array A with FasTensor. The convolution (denoted with *CONV_UDF*) has k kernels whose parameters are $w_{ij}, i \in [0, 8]$ and $j \in [0, k]$. The output of *CONV_UDF* is a vector, turning A into a 3D array. Then, we define the *POOL_UDF* and the *ReLU_UDF* on the data after the convolution. All three user-defined functions are chained together to be executed by calling the *Transform* function within the *main* function. The *main* function is the starting point for the execution of the whole program. Result data is stored into an array namely R

Similar to the previous study to detect AR events [88], we consider the layer close to the Earth's surface and the variables, namely *TMQ*. We use the output data for the year *1979*, which is a 768×1152 2D array. Apache Spark uses two sets of *Map* and *Reduce* functions to express *CONV* and *Pooling* layers, respectively. The *ReLU* is expressed with a *Map*. An implementation of CNN with FasTensor is shown in Fig. 2.4. Details about how to implementation these analysis steps in Apache Spark can be found in our previous work [7]. A performance comparison of

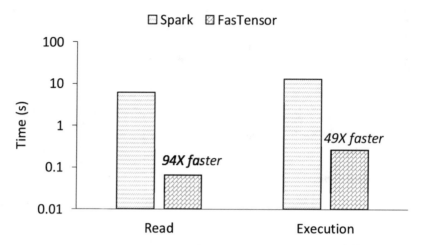

Fig. 2.5 Performance comparison of FasTensor with Spark for computing essential steps in CNN (CONV, Pooling and ReLU) on a 2D CAM5 dataset

FasTensor with Apache Spark is shown in Fig. 2.5. As expected, Spark's data read time is $\approx 94\times$ higher than that of FasTensor. The FasTensor framework reads multi-dimensional arrays without requiring to flatten them into key-value (KV) pairs, which gives the performance advantage to FasTensor. In computing the CNN steps ("Execution" stage in the Fig. 2.5), FasTensor is $49\times$ faster than Spark. FasTensor also benefits from reducing communication to synchronize only a small number of overlap layers with the preserved structural locality, where as Spark's linearization to a 1D KV causes significant communication overhead for the *Reduce* operation to gather inputs. As a result, FasTensor sees a high performance speedup compared to Apache Spark.

2.5.2 Gradient Calculation on a Plasma Physics Dataset

Vector Particle-in-cell (VPIC) simulates the magnetic reconnection phe-nomenon [40]. Studying of magnetic reconnection phenomenon using this data includes calculations of gradient on a 3D field mesh and finding gradient field value for each particle via interpolation. These operations involve four arrays, M, X, Y, and Z, where M is a 3D magnetic field mesh data and X, Y and Z contains particle locations. The gradient computing for M uses a Laplace operation, i.e., 3D version of the one in Fig. 2.6. Using Spark, the gradient on M is expressed with a *Map* and a *Reduce*, where the *Map* duplicates each cell for its neighbors and the *Reduce* operation performs the Laplace calculation. Then, a tri-linear operation is finished with a map-side join, where the gradient value of *M* is broadcasted to each executor and then a *Map* is used to find the gradient field value of each particle.

```
//User defined function on the input Stencil object iStencil from M for computing gradient
Gradient_UDF( iStencil )
{
        oStencil = 4iStencil_{0,0} − (iStencil_{−1,0} + iStencil_{0,−1} + iStencil_{0,1} + iStencil_{1,0}) ;
        return oStencil ;
}

//User defined function on the input Stencil iStencil from P for interpolation
//G contains the gradient of the space
Interpolate_UDF( iStencil ) {
{
        oStencil = BilinearInter(iStencil_{0}.X, iStencil_{0}.Y, G) ;
        return oStencil ;
}

main()
{
        G= Transform(M, Gradient_UDF) ;

        R = Transform(P, Interpolate_UDF) ;
}
```

Fig. 2.6 Sample code of gradient computing and interpolation functions with FasTensor on a 2D field data M and on a particle data P. The P has two location attributes X and Y. For simplicity, we use *BilinearInter* to denote the interpolation formula. In parallel execution, intermediate array G can be cached in memory and broadcast to all processes for performance. Note that we omit the *BilinearInter* function for simplicity and more detail about it can be found here: https://en.wikipedia.org/wiki/Bilinear_interpolation. R contains the result data

Implementation with FasTensor is a 3D version of the algorithm in Fig. 2.6. Since Spark has a limit on the size of broadcast data, we have set the test to use a small 256MB ($512 \times 256 \times 256$) field data. The particle data has 263GB with $\sim$ 23 billion particles. The tests used *128* CPU cores on 16 nodes.

In Fig. 2.7, we compare FasTensor's performance with that of Spark. We show that FasTensor is $106\times$ faster than Spark to execute the aforementioned gradient computations. In reading the input data, FasTensor is $45\times$ times faster than Spark. Because Spark has to duplicate a ($\sim 6\times$) cells to form the *reduce* operations to calculate the Laplace operator, its performance is poor. In contrast, FasTensor uses logical partitioning without duplication of data to finish the Laplace operator. Explicit processing and communication of array index for particle data in Spark also degrades its performance.

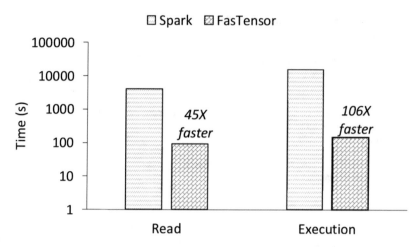

Fig. 2.7 Performance comparison of FasTensor and Spark for computing gradient using a plasma physics simulation (VPIC) dataset

2.6 Summary

In the same way as the MapReduce paradigm revolutionized data analysis for Key Value pair data model, we introduced FasTensor for analyzing multi-dimensional array data. This chapter introduced the main data structures of FasTensor, i.e., *Stencils* that can be expressed with chunks and overlap regions and *Transform* , the principle operator of FasTensor. The execution engine of FasTensor performs efficient and automatic parallelization of data movement, execution on multiple processes, and storing the final result on a given storage location. FasTensor also allows users specifying self-describing file formats, such as HDF5 and netCDF, as endpoints for reading the data for analysis and writing the end result into files. A performance evaluation of FasTensor with Apache Spark using a climate dataset and a plasma physics dataset demonstrates that FasTensor achieves superior performance in performing analysis functions such as convolution and gradient calculations.

Chapter 3
FasTensor User Interface

In this chapter, we describe the user interface of the FasTensor library, a C++ implementation of the FasTensor programming model presented in previous chapters.
Hence, the term *FasTensor* will mostly refer to the FasTensor library that implements the FasTensor programming model. Currently, the FasTensor programming
model is available in C++. Support for other languages, such as Java, Python, and
Julia, are planned.

This chapter is organized as follows. We present a simple but complete example
to introduce the basic programming style of FasTensor in Sect. 3.1. This example
serves as a good template for users to develop their own data analysis programs
with FasTensor. Following the example section, we provide a detailed description of
the FasTensor application programming interface (API) together with explanations
and coding examples. Most of these API example codes are ready to be compiled
and tried. Users are encouraged to install FasTensor and exercise these code
snippets; and the installation guide of FasTensor in Appendix A.1 gives more detail
instructions. Below are a few directions to read this chapter efficiently or to use the
contents of this chapter as analysis functions:

- **How to read this chapter?** The sections of this chapter is ordered to provide
  instructions for readers to construct their own analysis programs. It starts with
  a complete example in Sect. 3.1 and continues with descriptions of the core
  concepts represented by two C++ classes: **Stencil** in Sect. 3.2 and **Array** in
  Sect. 3.3, with additional auxiliary functions in Sect. 3.4. We encourage the first-
  time reader to go through the chapter without skipping sections. The sections
  are written to be self-contained so that experienced users may directly jump to a
  specific function or topic.

  The highlighted keywords in this chapter have the following conventions.

  - "File name", with the regular font and between quotation marks, refers to the
    name of a file that contains data or source code. This may also highlight a

© The Author(s), under exclusive license to Springer Nature Switzerland AG 2021   23
B. Dong et al., *User-Defined Tensor Data Analysis*, SpringerBriefs in Computer
Science, https://doi.org/10.1007/978-3-030-70750-7_3

directory name or a command (i.e., an executable file) from Linux shell, C++ compiler, HDF5 library, or compiled FasTensor programs.

- *"Dataset name"*, with the italic font and between quotation marks, refers to the a particular dataset name within a file. Following the HDF5 nomenclature, we use dataset to refer to a multidimensional array. In an ADIOS file, a dataset is a variable referring to a multidimensional array.
- **class name**, with the boldface font, refers to the C++ class name in the FasTensor library.
- *class member*, with the italic font (without quotation marks), refers to method names or attributes of an C++ class in FasTensor. We also present standalone function names in this style.
- ***parameter name***, with both italic and boldface font, refers to the argument of a method of the C++ class or a function.

- **How to find the example code snippets?** When the FasTensor source code is downloaded and unpacked, all example code can be found in a sub-directory named "tutorial". Since the source code of FasTensor is developed with the 2017 C++ standard, it is necessary to instruct your compiler to include C++17 features, for example, by specifying the following option -std=c++17. The "tutorial" directory has a simple "Makefile" for users to compile all the examples, but users may need to modify "Makefile" to direct compilers on their systems and to accommodate other system-specific settings. Users can find and compile the tutorial code with the following command.

```
$ cd tutorial # from the root of  FasTensor
      source code
$ make
```

- **How to generate the data used in the examples?** In the "tutorial" sub-directory within the FasTensor source code tree, we provide a tool named "create_data.cpp" to generate tutorial data sets used later in this chapter. For example, the command line in the Listing 3.1 generates a 2D 16 × 12 dataset of float type.

Listing 3.1 An example to create a sample data set in a HDF5 file and to display the file content to console

```
1    $ ./create_data -f ./tutorial.h5 -d /dat -n 2 -s 16,12
2    $ h5dump  tutorial.h5
3      HDF5 "tutorial.h5" {
4        GROUP "/" {
5          DATASET "dat" {
6            DATATYPE  H5T_IEEE_F32LE
7            DATASPACE  SIMPLE { ( 16, 12 ) / ( 16, 12 ) }
8            DATA {
9            (0,0):     1.81,    2.81, ... ,  11.81,   12.81,
10           (1,0):    13.81,   14.81, ... ,  23.81,   24.81,

11            :         :        :              :        :
12           (14,0): 169.81, 170.81, ... , 179.81, 180.81,
13           (15,0): 181.81, 182.81, ... , 191.81, 192.81
14    }}}}
```

Listing 3.1 contains two commands in Lines 1 and 2 following the command line prompt $. The first line runs the executable "create_data" compiled from "create_data.cpp" to create a sample data file and the second line invokes HDF5 tool "h5dump": [1] to display the content of the file generated in Line 1. The program "create_data" accepts four options: "-f" for the file name, "-d" is the dataset name, "-n" for number of dimensions, and "-s" for dataset size. From Listing 3.1 above, "create_data" produces a HDF5 file named "./tutorial.h5" with a single dataset named "*/dat*". This dataset is a 2D array, which has 16 elements in the first dimension and 12 elements in the second dimension.

From the output of "h5dump", we can see that the dataset name, data type, and array dimensions are as specified. For those who are unfamiliar with HDF5 terminology, what is common referred to simply as float is more precisely described as H5T_IEEE_F32LE at the Line 6, and the simple 2D array is termed "DATASPACE SIMPLE"[2] in Line 7 of Listing 3.1. The content of the dataset "*/dat*" in file "./tutorial.h5" goes from 1.81 to 192.81.

- **Which file format is used in the examples?** FasTensor supports a number of file formats, including HDF5,[3] ADIOS,[4] PNetCDF,[5] TDMS,[6] and Binary. Our example code in this book assumes that all data are stored within HDF5 files since it is the default file format for FasTensor. The protocol to add another file format to FasTensor is described in Appendix A.2 at the end of this book.

3.1 An Simple Example of Using FasTensor

In this section, we present a simple example to illustrate the usage of FasTensor. This example processes a data set stored in a file namely "tutorial.h5". The file "tutorial.h5" is created by commands in line 1 of Listing 3.1. The dataset namely "*/dat*" in the file "tutorial.h5" is a 16 by 12 2D array. To make it meaningful in the context of an application, we assume that each row of this 2D array is a time series (with 12 points) from a sensor, and 16 rows represent 16 sensors. In Listing 3.2, we present a sample source code to process a data file stored in the HDF5 format (i.e., "tutorial.h5") using a three-point moving average operation in FasTensor.

Note that the program code in Listing 3.2 can run on a single CPU core (i.e., sequential) or on multiple CPU cores or multiple computing nodes (i.e., parallel).

---

[1]"h5dump" https://portal.hdfgroup.org/display/HDF5/h5dump

[2]Because HDF5 supports variable sized arrays, the extend of each dimension is given by two numbers, the current size and the maximum size. In this example, the current size and the maximum size of each dimension are all 16 by 12.

[3]https://www.hdfgroup.org/

[4]https://www.olcf.ornl.gov/center-projects/adios/

[5]https://parallel-netcdf.github.io/

[6]https://www.ni.com/en-us/support/documentation/supplemental/06/the-ni-tdms-file-format.html

Fig. 3.1 An example of endpoint specification for an array stored as a HDF5 dataset in a HDF5 formatted file

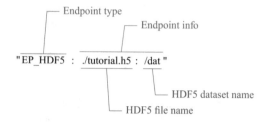

To run on a single core, one can run with a single MPI process. The program code has a typical structure as most normal C++ programs. It has the *main* function (in Line 10) as the starting point of the program and then it executes the following parts line by line. Detailed explanations of these lines are presented in the following paragraphs. We provide instructions to compile and run this code at the end of this section.

In line 13 of Listing 3.2, the example code calls the *FT_Init* function from FasTensor to initialize the FasTensor environment. The function *FT_Init* initializes the Message Passing Interface (MPI) runtime[7] and other dependencies. *FT_Init* takes arguments from the *main* function, i.e., **argc** and **argv**. This initialization call is required before any other functions from FasTensor can be executed. Before the end of *main*, in Line 30 of the program, *FT_Finalize()* must be called to release resources used by FasTensor. Please see details of the resource release in later Sect. 3.4.2. All these functions from FasTensor are contained in a namespace "FT". Details of *FT_Init* and *FT_Finalize* are presented in Sect. 3.4.

In line 19, we show FasTensor's using an in-memory **Array** object A to refer to the input data. In line 21, it uses another in-memory **Array** object B to represent the output data. In this example, both the object A and the object B point to two 2D arrays that are stored as HDF5 datasets in files, separately. We call these data sets as "endpoints" in FasTensor. By taking a parameter for object A as an example, i.e., "EP_HDF5:./tutorial.h5:/dat" is an endpoint identification. It contains two parts that are separated by a 'colon' symbol. The first part of this endpoint description is the type of an endpoint. In this example, the endpoint type is EP_HDF5, which means that data variables are stored in a HDF5 file. Then, the second part contains a string "./tutorial.h5:/dat", which represents the "endpoint information" or "endpoint info" as shown in Fig. 3.1. The endpoint info has two strings separated by a 'colon' symbol again. The first part, i.e., "./tutorial.h5", specifies the path of a HDF5 file name. The second part, i.e., "/dat" specifies a HDF5 dataset or array variable path and its name within the HDF5 file. In summary, this format for initialization string for a HDF5 file dataset is presented in Fig. 3.1. The Table 3.1 in the following section introduces the format for other files such as ADIOS or PNetCDF.

In line 19, **Array** object A also has two other parameters, **chunk size** and **overlap size**. The **chunk size** is a 1D C++ vector that has values of {4,4}. The **chunk size** indicates that Array object A is going to be split into subsets and each

---

[7]MPI is designed to be a process manager and communication protocol for programming parallel computers. Learn more about MPI at https://www.mpi-forum.org/.

Listing 3.2 An example to use FasTensor to perform a three-point moving average operation on a 2D array stored in a HDF5 file. The three-point moving average operation is expressed as a user-defined function *udf_ma*. FasTensor uses the C++ *main* function to set up both input and output, to transform the data from the input to the output based on the expression in *udf_ma*. Note that the example code can run either sequentially on a single CPU or parallel on multiple CPUs across multiple computing nodes without any modification. The symbol ↪ serves as a "line break" in a long line of code

```
1   #include "ft.h"
2
3   inline Stencil<float> udf_ma(const Stencil<float> &iStencil)
4   {
5       Stencil<float> oStencil;
6       oStencil = (iStencil(0,-1) + iStencil(0,0) +
                ↪iStencil(0,1))/3.0;
7       return oStencil;
8   }
9
10  int main(int argc, char *argv[])    //start here
11  {
12      //Initialize FasTensor , MPI, etc.
13      FT::FT_Init(argc, argv);
14
15      //Set up chunk size and overlap size
16      std::vector<int> chunk_size = {4, 4};
17      std::vector<int> overlap_size = {0, 1};
18      //Input data
19      FT::Array<float> *A = new
                ↪FT::Array<float>("EP_HDF5:./tutorial.h5:/dat",
                ↪chunk_size, overlap_size);
20      //Result data
21      FT::Array<float> *B = new
                ↪FT::Array<float>("EP_HDF5:./tutorial_ma.h5:/dat");
22      //Run
23      A->Transform(udf_ma, B);
24      //Create Vis file for ParaView or VisIt
25      B->CreateVisFile();
26      //Clear
27      delete A;
28      delete B;
29
30      FT::FT_Finalize();
31
32      return 0;
33  }
```

has the size as 4 by 4. The ***overlap size*** is also a 1D C++ vector that has values of {0, 1}. In this example, the ***overlap size*** makes each chunk to be extended by 1 point in the second dimension. Combining these two parameters, the HDF5 dataset "*/dat*" is broken into 12 subsets and each has the size of 4x6 (except the chunks containing points out of boundary). Each chunk is processed independently from the other subsets in FasTensor and subsets are iteratively processed one by one in FasTensor. Combining with the ***chunk size*** and ***overlap size***, the running of three-point moving average can happen either sequentially or in parallel on different subsets. Also, these parameters ensure that there is no communication occurring during the execution of the program.

In line 21 of Listing 3.2, the output Array object B is initialized with an "EP_HDF5" endpoint. We note that the endpoint type for Array B can be different from that of A. Users can specify any supported type of endpoint for the output array B. These is no need to specify the ***chunk size*** and the ***overlap size*** for the object B in this example because array B will inherit the ***chunk size*** and the ***overlap size*** from the object A. The output data for object B will be directly written to another HDF5 file, named "./tutorial_ma.h5", in this example. FasTensor can figure out the storage location automatically for storing the output data from processing array A. FasTensor can store the data using either sequential or parallel I/O operations. If users want to store array B for other operations after the three-point moving average operation, users can declare array B with an in-memory endpoint whose type code is "EP_MEMORY". The "EP_MEMORY" endpoint can cache the output data from A in memory.

Now that we have the basic settings for the input data and the output data, it's time to explain how users can express the three-point moving average operation itself. FasTensor provides a user-defined function (UDF) mechanism and a Stencil abstraction to easily customize the three-point moving average operation. The Stencil abstraction is implemented as the **Stencil** class. More details about the **Stencil** class are in the following Sect. 3.2. In this example, the user-defined code for the moving average operation is in function *udf_ma*, which is between line 3 and line 8 in Listing 3.2. This function takes objects of the **Stencil** class as both input and output. As mentioned earlier, the **Stencil** class is defined by FasTensor to represent a logical neighborhood of an array and therefore to express the UDF operation on each neighborhood cell. Hence, the **Stencil** class gives a natural way for a user to customize three-point moving average because it needs three neighbors from a neighborhood to perform the average calculation. In the following paragraph, we describe how the **Stencil** class works in the context of the three-point moving average.

In this example, since the moving average needs three points to calculate average, the neighborhood for each operation only needs to contain three values. The **Stencil** class has overloaded the parenthesis operator "()" to allow users to access each neighborhood cell by their relative offset, as shown by line 6. The cell at iStencil(0,0) is special one, which accesses the base cell of the neighborhood. The base cell is the place where the UDF operation happens. The expression iStencil(0,-1) accesses the point before the base cell and iStencil(0,1) accesses the point after the base cell. This is shown in the left side of Fig. 3.2. The overloaded operator "()"

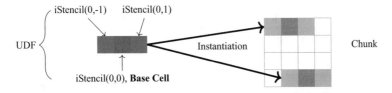

Fig. 3.2 A user-defined function (UDF) describes the operation on the base cell and it neighbors. The described operation is instantiated on all cells by FasTensor with its Transform function

Listing 3.3 Compile and run the example code with the three-point moving average operation

```
1    # ft_ma.cpp is the file containing the example code
2    # compile it with g++ and flags from FasTensor ,
3    # MPI, HDF5 and other system specific ones.
4    $ g++ ft_ma.cpp   -o ft_ma    "other flags"
5
6    # or compile it with mpic++ and
7    # other flags from FasTensor  and HDF5
8    $ mpic++ ft_ma.cpp   -o ft_ma   "other flags"
9
10   #run the code sequentially
11   $ ./ft_ma
12
13   #or   run the code in parallel
14   $ mpirun -n 4 ./ft_ma
```

returns the value which users can directly use to perform moving average in line 6. The calculated result is assigned to the output **Stencil** object, namely oStencil. The oStencil object serves as the return value at the end.

Let us now combine all above together to see how the data analysis expression (as UDF[8]) gets executed onto the input **Array** object A. In line 23 of Listing 3.2, FasTensor runs the UDF *udf_ma* through the *Transform* function of the Array object A. Intuitively, the *Transform* function executes the *udf_ma* function on each array cell from the A. Internally, *Transform* instantiates an input **Stencil** object "iStencil" for each cell in the A (as illustrated in Fig. 3.2), feeds the "iStencil" object into *udf_ma*, and extracts the output value from the output **Stencil** object, i.e., "oStencil". Also, the *Transform* function writes the output data into the Array object B. If users run the code in parallel using MPI, the *Transform* function also automatically deals with other underlying tasks, such as the assignment of chunks among processes. Users can compile the code in Listing 3.2 once and then run it either as a sequential program or a parallel program without any further modification.

---

[8]UDF: user-defined function. In C++, the UDF is the type of normal function defined by users. See more details at https://www.cplusplus.com/doc/tutorial/functions/.

Listing 3.4 Examine the result file "tutorial_ma.h5" with the "h5dump" command

```
1      $ h5dump  tutorial_ma.h5
2         HDF5 "tutorial_ma.h5" {
3         GROUP "/" {
4            DATASET "dat" {
5            DATATYPE  H5T_IEEE_F32LE
6            DATASPACE  SIMPLE { ( 16, 12 ) / ( 16, 12 ) }
7            DATA {
8               (0,0):  2.14333,   2.81, ... ,  11.81, 12.4767,
9               (1,0):  14.1433,  14.81, ... ,  23.81, 24.4767,

10                           :         :    :      :       :

11              (14,0): 170.143, 170.81, ... , 179.81, 180.477,
12              (15,0): 182.143, 182.81, ... , 191.81, 192.477
13      }}}}
```

Finally, we can move on to compiling and running the program. As mentioned at the beginning of this section, FasTensor uses MPI by default for parallel execution. Hence, users can run the compiled source code as a normal MPI program. Specifically, assuming that a user has installed FasTensor and its dependencies successfully, one can compile and run the example code with the commands shown in Listing 3.3. Let's assume that the example source code is stored in a file, named "ft_ma.cpp". Users can compile the code with either "g++" (line 4) or "mpic++" (line 8) with proper compiler flags. Then, users can run it as sequential code with a single process (i.e., 1 CPU core) line 11. If one has a system with multiple CPU cores or that with multiple nodes, it can be executed in parallel mode with the MPI command "mpirun" (shown in line 14). Depending on the MPI installation or scheduler on a system, we would note that the user may have to replace "mpirun" with "mpiexec" or "srun" or "jsrun". The "-n 4" argument in the "mpirun" command conveys that MPI uses four processes to run the code. Given the configuration of the computer system and MPI, these four MPI processes can started on a single computing node or multiple computing nodes.

One may recall that the output file is stored in another HDF5 file, named "tutorial_ma.h5" (pointed by the **Array** object B in line 21). Users can examine the content of the "tutorial_ma.h5", as shown in Listing 3.4. We also use a HDF5 tool called "h5dump", to show the contents of the array in ASCI format. One may use HDF5 viewer tools (e.g., h5view) to visualize the contents of HDF5 files. The dataset "/dat" in "tutorial_ma.h5" has the same size (16×12) as the input file, "tutorial.h5". The element type of the output array is 'float' ("H5T_IEEE_F32LE"). Note that the point on the boundary has different calculation because of the padding effect. By default, FasTensor duplicates the nearest boundary points for out-of-boundary points during the run of the UDF *udf_ma*. Say for the left-top most point 2.14333, it is calculated from $\frac{1.81+1.81+2.81}{3}$, where the first 1.81 is a duplicated value of the 1.81 at index (0,0) of the input array. The second 1.81 is the value at index (0,0) of the input array too. The 2.81 is the value at the index (0,1) of the input

Fig. 3.3 A view of the result file "tutorial_ma.h5" via plotting it in the ParaView

array. Other options are also available in FasTensor to deal with the boundary point, as described by the following Sect. 3.3.3.

It is worth noting that FasTensor can also help users to generate some visualization file on their data. In line 25, the B object calls its method *CreateVisFile* to create a XDMF file (namely "tutorial_ma.h5.xdmf") to plot the "tutorial_ma.h5" in visualization software, such as ParaView and VisIt. The "tutorial_ma.h5.xdmf" file is created within the same directory as the "tutorial_ma.h5". The Fig. 3.3 shows the results produced by ParaView with the XDMF file. More details about the visualization file can be found in the following Sect. 3.3.13.

3.2   The Stencil Class

The philosophy of FasTensor is to break a complex data analysis operation into lots of small and self-contained units. These units can be easily understood and customized by users. Also, these units can also be accessed, scheduled and executed independently. One important conception in supporting this idea is the **Stencil** class, an abstract for the neighborhood of a multidimensional array. The **Stencil** class implements the new neighborhood abstract in FasTensor for the multidimensional array data. Conceptually, a **Stencil** object has a base cell, where FasTensor runs the data analysis operation from the user-defined function. The **Stencil** object can access its neighborhood cell at any place using relative offsets from the current base cell. As shown in the previous section, the object of the **Stencil** class is used as both the input and the output of the user-defined function (UDF) of FasTensor. The UDF is the place where users write the data analysis code. Hence, FasTensor requires users to format the input and output of their data analysis code as **Stencil** objects.

In Listing 3.5, we summarize the major public methods from the **Stencil** class within FasTensor. These methods can help users to format the input and the output for the data analysis code. Note that the Listing 3.5 provides pseudocode-like implementation of the **Stencil**. In the real source code of the **Stencil** class in FasTensor, it may have minor difference in parameter names or have extra C++ keywords like *const* to make them more efficient. Of course, these differences in parameter names and C++ keywords will not impact the usage and the signature of these methods. Basically, these parameter names or keywords are just the place holder or the optimization hint in the C++ programming language.

For the sake of simplicity, we ignore other internally used (private or protected) functions from the **Stencil** class. For example, during the explanations of the example code in the Listing 3.2 in the previous section, the *Transform* function of FasTensor creates a **Stencil** object for each cell of an array, feeds these objects (as input) into the user-defined function during the execution, and extracts the output value from the output **Stencil** objects. Since these operations on **Stencil** are transparent to users, we ignore them in this section. In other words, we only report the methods of **Stencil** that users can call and use in their application codes.

The **Stencil** class accepts a template parameter T to indicate the type of the data element that the **Stencil** object will carry. Theoretically, the **Stencil** object can be created with any type that is supported by the C++ language. The **Stencil** class has three constructor functions. The **Stencil** class also overloads the operator parenthesis "()" and the operator equal "=". The overloaded operator parenthesis "()" is mostly used to access a single neighborhood cell at a relative offset. The equal "=" operator is mostly used to assign output value as the return value of a user-defined function. A generic version of the operator parenthesis "()" is also provided as *ReadNeighbors* and *WriteNeighbors* to access a contiguous set of neighborhood cells. The *SetShape* function is used to specify the shape of the data that are represented by the **Stencil** object. The *SetShape* function is mostly used on the output **Stencil** object whose data type T is a vector.

The **Stencil** class also provides the *GetMaxOffsetUpper* function and the *GetMaxOffsetLower* function for users to get the maximum offset at either the upper side or the lower side. The *GetChunkID* method is used to get the chunk identification of the chunk where the current **Stencil** created from. The *GetGlobalIndex* method and *GetLocalIndex* method are used to obtain the coordinate of the base cell within either the whole array or the chunk. The *GetChunkID* method, the *GetGlobalIndex* method and the *GetLocalIndex* method can help users to pivot the base cell as well as the **Stencil** object during the execution of the *Transform* function. Details of the these methods are presented in the following parts of this chapter.

Listing 3.5 An overview of the **Stencil** class from FasTensor and its major methods for application users. Note that the symbol "$" denotes the output parameters for the methods which have output value. In the C++ language, the symbol "$" is used to create an alias for a reference variable. Hence, users must create an object for the output parameter to store the output value

```
1    template <class T>
2    class Stencil{
3    public:
4      Stencil();
5      //Constructor with  value
6      Stencil(T value);
7      //Constructor with value and its shape
8      Stencil(T value, vector<size_t> shape);
9      //read a neighbor at relative offset (ri, rj, ....)
10     T operator()(int ri, int rj, ...);
11     //A vectorized version of operator () at relative offset (ro)
12     T ReadPoint(vector<int> ro)
13     //set value
14     void operator=(T value);
15     //read contiguous neighbors from start_ro to the end_ro
16     int ReadNeighbors(vector<int> start_ro, vector<int> end_ro,
                ↪vector<T> &data);
17     //write contiguous neighbors from start_ro to the end_ro
18     int WriteNeighbors(vector<int> start_ro, vector<int> end_ro,
                ↪vector<T> data);
19     //set the shape of the output vector
20     int SetShape(vector<size_t> shape);
21     //get the shape of the output vector
22     int GetShape(vector<size_t> &shape);
23     //set the value
24     int SetValue(T value);
25     //get the value
26     int GetValue(T &value);
27     //Get maximum offset at upper side
28     int GetOffsetUpper(vector<int> &upper);
29     //Get maximum offset at lower side
30     int GetOffsetLower(vector<int> &lower);
31     //Get chunk id
32     int GetChunkID(unsigned long long &chunk_id);
33     //Get index of the base cell within the array
34     int GetGlobalIndex(std::vector<unsigned long long> &index);
35     //Get index of the base cell within the chunk
36     int GetLocalIndex(std::vector<unsigned long long> &index);
37   };
```

3.2.1 Constructors of the Stencil Class

Signature:

```
Stencil();
Stencil(T value);
Stencil(T value, vector<size_t> shape);
```

Description:

The **Stencil** class has three constructors for users to build a **Stencil** object. The **Stencil** object is mostly used as the return (output) value of an user-defined function because FasTensor automatically builds the input **Stencil** object transparently. The first constructor has no parameter and the second one accepts a parameter *value*. The last one accepts two parameters, *value* and *shape*.

Parameter:

- *value* contains the value for the Stencil object to build. The type of value *T* can be most fundamental types (e.g., *float* or *double*) from C++ or the *vector* type from the compound type of C++.
- *shape* is a vector with the size n to specify the size for the neighborhood represented by the **Stencil** object to build. It is only needed when one has 1D vector (e.g., std::vector) as the output type *T*. In other words, the 1D vector is actually flatted from a subset from a n-dimensional array and the subset has the size of *shape*.

Example:

The Listing 3.6 gives three examples to show the usage of three **Stencil** constructors. All examples assume that the input data is a 2D array. The first UDF *UDF_SC1* and the second UDF *UDF_SC2* multiple each point by 2 and thus the returned **Stencil** object only have one value. The first UDF *UDF_SC1* uses the *SetValue* method to assign value to the object which is built from the constructor *Stencil()*. More details about the *SetValue* method can be found in Sect. 3.2.4. The second UDF *UDF_SC2* directly uses the constructor *Stencil(T value)*. The third UDF *UDF_SC3* returns a 1D vector and it has the shape size as 4 by 4. Usually, the *UDF_SC3* is used jointly with the stride operation of the **Array** of FasTensor (see details in the Sect. 3.3.5 of the following chapter) to process the data.

3.2.2 Parenthesis Operator () and ReadPoint

Signature:

```
T operator()(int ri, int rj, ...);
T ReadPoint(vector<int> ro)
```

Listing 3.6 Example of using Stencil constructors to build output objects for three different user-defined functions. For the sake of simplicity, we omit the "("main) function in the example

```
//Example 1: use the Stencil constructor to create a return
//object with a single value, from the input Stencil object
Stencil<float> UDF_SC1(const Stencil<float> &iStencil){
  Stencil<float> oStencil;
  oStencil.SetValue(iStencil(0,0) * 2);
  return oStencil;
}

//Example 2: use the Stencil constructor to create a return
//object with a single value, from the input Stencil object
Stencil<float> UDF_SC2(const Stencil<float> &iStencil){
  return Stencil<float> (iStencil(0,0) * 2);
}

//Example 3: use the Stencil constructor to create a return
//object containing a 4 by 4 array subset (flatted)
Stencil<vector<float>> UDF_SC3(const Stencil<float> &iStencil){
  std:vector<float> output_vector(16)
  for(int i= 0;  i < 4 ; i++){
    for(int j= 0;  j < 4 ; j++){
      output_vector[i*4+ j] = iStencil(i, j) * 2;
  }}
  vector<size_t> shape = {4,4};
  return Stencil<vector<float>>(output_vector, shape);
}
```

Description:

The parenthesis operator () of the Stencil class receives the relative address of a neighbor from the base cell and returns its value. The return value has the same template type **T** as the Stencil class. The *ReadPoint* method has the same function as the operator (). The *ReadPoint* function can be used when users want to use the relative_offset (**ro**) vector to contain the relative address, e.g., **ri** and **rj**.

Parameter:

- **ri, rj, ...** are the relative address of a neighboring array element to access. Both can be positive and negative values. The three dots ... in the API means that given a n-dimensional data, users can specify at most n address.
- **ro** is the relative_offset vector with the size n. The $ro[i]$ is the relative address of a neighbor at the ith dimension of an array.

Example:

Listing 3.7 gives an example to show how to use the operator () from the **Stencil** class. This example performs the convolution computing via accessing four neighbors from each base cell. The convolution computing consists of three kernel filters and each has the size of 2 by 2. We fill some random values as the kernel

Listing 3.7 An example of using the parenthesis operator () to express the convolution computing on a 2D array data. The convolution has 3 kernels and each has the size 2 by 2. The user-defined function *UDF_Cov* can be called by the *Transform* function of the **Array** class to execute on the whole array. FasTensor also transparently creates the output data which is a 3D array

```
#define KERNELS 3
#define KERNEL_HEIGHT 2
#define KERNEL_WIDTH 2
float kernel_weight[KERNELS][KERNEL_HEIGHT][KERNEL_WIDTH] =
{
    {{1, 2}, {3, 4}},
    {{2, 3}, {4, 5}},
    {{3, 4}, {5, 6}},
};
Stencil<vector<float>> UDF_Cov(const Stencil<float> &iStencil)
{
    float temp_sum = 0;
    vector<float> conv_vec;
    for (int k = 0; k < KERNELS; k++)
    {
        for (int i = 0; i < KERNEL_HEIGHT; i++)
        {
            for (int j = 0; j < KERNEL_WIDTH; j++)
            {
                temp_sum += iStencil(i, j) *
                    ↪kernel_weight[k][i][j];
            }
        }
        conv_vec.push_back(temp_sum);
        temp_sum = 0;
    }
    std::vector<size_t> shape = {1, 1, KERNELS}
    return Stencil<vector<float>>(conv_vec, shape);
    //One can replace the above line with the below lines
    //  Stencil<vector<float>> oStencil();
    //  oStencil.SetValue(conv_vec);
    //  oStencil.SetShape(shape);
    //  return oStencil;
}
```

weights and apply them on 2 by 2 array subsets, which are represented by the iStencil object. The output of this operation is a 1D vector with the length of three. Then, we set the shape of the output **Stencil** object to let FasTensor handle the output vector correctly. Basically, the output from a convolution on the 2D array is a 3D array. So, each output of the convolution is a 3D array subset. The shape of the output **Stencil** object is 1 by 1 by 3 in this case. More details about the shape of the output vector are presented in the following subsection.

3.2.3   SetShape and GetShape

Signature:

```
int SetShape(std::vector<size_t> shape);
int GetShape(std::vector<size_t> &shape);
```

Description:

The *SetShape* function sets the **shape** of a **Stencil** object which contains a 1D vector typed value (i.g., *std::vector*). Note that the **Stencil** object is mostly used as the return value of an user-defined function. In this case, the user-defined function should linearize the data properly based on the **shape**. By default, the row-major data layout is assumed by FasTensor to handle the data contained in the output Stencil object. Basically, the **shape** describes the size of the subset of the output array. The *GetShape* method returns the **shape** of the Stencil object. The *GetShape* method is mostly used by FasTensor to obtain its **shape** parameter internally.

Parameter:

- *shape* is a vector with the size n that specifies the size of a subset from the output array where this Stencil object will be stored. The *shape[i]* is the size for the ith dimension of a multidimensional array.

Example:

Please see Listing 3.7 for the example code in the previous subsection about the usage of *SetShape*.

3.2.4   SetValue and GetValue

Signature:

```
int SetValue(T value);
int GetValue(T &value);
```

Description:

The *SetValue* function assigns **value** to a **Stencil** object, which is mostly used as the return value of an user-defined function. The *GetValue* method returns the value from the **Stencil** object. The template type T is the same as the one for the **Stencil** class.

Parameter:

- *value* contains the value to be assigned during the call of the *SetValue* function and the value from the **Stencil** object during the call of the *GetValue* function.

Example:

Please see the Listing 3.7 for the example code in the previous subsection about the usage of *SetValue* method.

3.2.5   ReadNeighbors and WriteNeighbors

Signature:

```
int ReadNeighbors(vector<int> start_ro, vector<int> end_ro,
    ↪vector<T> &data);
int WriteNeighbors(vector<int> start_ro, vector<int> end_ro,
    ↪vector<T> data);
```

Description:

The *ReadNeighbors* function and the *WriteNeighbors* function access values from neighborhood in a batch and contiguous manner. Users can specify ***start_ro*** (i.e., starting relative offset) and ***end_ro*** (i.e., ending relative offset). Both ***start_ro*** and ***end_ro*** are relative offsets from the base cell. The *ReadNeighbors* function returns the value from ***start_ro*** to ***end_ro*** as a 1D vector (flatted into row-major order). The *WriteNeighbors* function writes the data (flatted into a 1D vector in row-major order) to the neighborhood from ***start_ro*** to ***end_ro***. An example to show the row-major order is presented in Fig. 3.4.

Parameter:

- ***start_ro*** is a vector with the size *n* that specifies the starting relative offset from the base cell of the **Stencil** object.
- ***end_ro*** is a vector with the size *n* that specifies the ending relative offset from the base cell of the **Stencil** object.
- ***data*** contains the data to write during the call of the *WriteNeighbors* function or the data to read by the *ReadNeighbors* function. Users must linearize their data into 1D vector before the write. The returned data from read are linearized by FasTensor too. By default, FasTensor uses the row-major ordering to linearize the data.

Example:

Listing 3.8 presents an example to use the *ReadNeighbors* function to access a contiguous neighborhood. This example uses *ReadNeighbors* to access four cells (2 by 2) to perform convolution operation. Comparing with the example code in

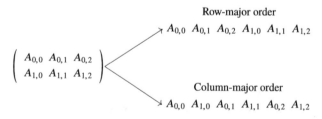

Fig. 3.4 Row-major order and column-major order for a 2 by 3 2D data. FasTensor assumes that all data are stored, accessed, and linearized into the row-major order

Listing 3.8 An example of using the *ReadNeighbors* function from the Stencil class

```
#define KERNELS 3
#define KERNEL_HEIGHT 2
#define KERNEL_WIDTH 2
float kernel_weight[KERNELS][KERNEL_HEIGHT][KERNEL_WIDTH] = {
    {{1, 2}, {3, 4}}, {{2, 3}, {4, 5}}, {{3, 4}, {5, 6}},
};
Stencil<vector<float>> UDF_Cov2(const Stencil<float> &iStencil)
{
    vector<int> sro = {0, 0}, ero = {1, 1};
    //Read s(0,1)  s(0,1) s(1,0)   s(1,1)
    vector<float> val_vec;
    iStencil.ReadNeighbors(sro, ero, val_vec);
    float temp_sum = 0;
    vector<float> conv_vec;
    for (int k = 0; k < KERNELS; k++)
    {
        for (int i = 0; i < KERNEL_HEIGHT; i++)
        {
            for (int j = 0; j < KERNEL_WIDTH; j++)
            {
                temp_sum += val_vec[i * KERNEL_WIDTH + j] *
                    ↪kernel_weight[k][i][j];
            }
        }
        conv_vec.push_back(temp_sum);
        temp_sum = 0;
    }
    std::vector<size_t> shape = {1, 1, KERNELS}
    return Stencil<vector<float>>(conv_vec, shape);
}
```

the previous Listing 3.7 that uses the parenthesis operator (), it only uses one *ReadNeighbors* call to get all needed cells from the neighborhood.

3.2.6   GetOffsetUpper and GetOffsetLower

Signature:

```
int GetOffsetUpper(vector<int> &upper);
int GetOffsetLower(vector<int> &lower);
```

Description:

These two functions return the maximum offsets that can be called from the base cell of the **Stencil** object. The *GetOffsetUpper* function returns the maximum offset

Listing 3.9 An example of using the *GetOffsetUpper* method and the *GetOffsetLower* method from the Stencil class

```
Stencil<vector<float>> UDF_MaxOffset(const Stencil<float>
    ↪ &iStencil)
{
  vector<int> offset_upper, offset_lower;
  iStencil.GetOffsetUpper(offset_upper);
  iStencil.GetOffsetLower(offset_lower);
  vector<float> output_vec;
  for (int i = offset_lower[0]; i <= offset_upper[0]; i++)
  {
    for (int j = offset_lower[1]; j <= offset_upper[1]; j++)
    {
      output_vec.push_back(iStencil(i,j) * 2);
    }
  }
  vector<size_t> shape(2);
  shape[0] = offset_upper[0] - offset_lower[0] + 1;
  shape[1] = offset_upper[1] - offset_lower[1] + 1;
  return Stencil<vector<float>>(output_vec, shape);
}
```

at the upper side. The *GetOffsetLower* function returns the maximum offset at the lower side.

Parameter:

- ***upper*** is a vector with the size n, which contains the maximum relative offset from the base cell to the upper boundary of the current Stencil object. The ***upper[i]*** contains the maximum offset at the ith dimension of an array data. In most cases, the upper boundary is calculated from the size of a chunk.
- ***lower*** is a vector with the size n, which contains the maximum relative offset from the base cell to the lower boundary of the current Stencil object. The ***lower[i]*** contains the maximum offset at the ith dimension of an array data. In most cases, the lower boundary is zero based vector.

Example:

Listing 3.9 gives an example to use the *GetOffsetUpper* function and the *GetOffsetLower* function to obtain the boundary of current **Stencil** object. Then, the example multiples each value by 2. The result value is stored into a 2D vector. The 2D vector is used as the return value of the output **Stencil** object. Meanwhile, it uses the ***shape*** parameter to set the size of the output data in the output array. Note that this example can be made more sense if it uses with the *SetStride* method from the **Array** class of FasTensor. More details about the *SetStride* method are presented in the following Sect. 3.3.5.

Listing 3.10 Example of using the *GetChunkID* to multiple each cell by the chunk ID on a 2D data

```
Stencil<float> UDF_DiagonalZero(const Stencil<float> &iStencil)
{
  unsigned long long chunk_id;
  iStencil.GetChunkID(chunk_id);
  return Stencil<float>(iStencil(0,0) * chunk_id)
}
```

3.2.7 GetChunkID

Signature:

```
int GetChunkID(unsigned long long &chunk_id);
```

Description:

This function return the ID of the chunk where the **Stencil** object created from. More details about the chunk and its ID are presented in the later Sect. 3.3.2. All the cells of a single chunk have the same chunk ID.

Parameter:

- *chunk_id* is the output parameter which contains the integer typed ID for the chunk of the calling **Stencil** object.

Example:

Listing 3.10 gives an example to multiple each cell by the chunk ID.

3.2.8 GetGlobalIndex and GetLocalIndex

Signature:

```
int GetGlobalIndex(vector<unsigned long long> &index);
int GetLocalIndex(vector<unsigned long long> &index);
```

Description:

The *GetGlobalIndex* function returns the index of the base cell in the global array. The *GetLocalIndex* returns the index of the base cell within the local chunk.

Parameter:

- *index* is a vector with n elements. The *index[i]* contain the coordinate for the base cell at the ith dimension.

Listing 3.11 An example of using the *GetGlobalIndex* method to convert all diagonal cells to be zero within a 2D array data

```
Stencil<float> UDF_DiagonalZero(const Stencil<float> &iStencil)
{
    vector<unsigned long long> index_global;
    iStencil.GetGlobalIndex(index_global);
    if(index_global[0] == index_global[1] ){
        return Stencil<float>(0);
    }else{
        return Stencil<float>(iStencil(0,0));
    }
}
```

Example:

Listing 3.11 gives an example to set all diagonal cells to be zero with the *GetGlobalIndex* function.

3.2.9   Exercise of the Stencil Class

This section presents an exercise to implement the convolution calculation with the operator () and the *ReadNeighbors* function. The example for these implementations are presented in Listing 3.7 and also in Listing 3.8. Users can refer to the example code in Listing 3.2 from the previous section about how to have a full implementation of this exercise. After implementation, we measure the time to run the user-defined convolution calculation on the "tutorial.h5" file. The "tutorial.h5" file contains a 16 by 12 2D array. The time that are obtained by authors is reported in the Fig. 3.5. Since FasTensor hides all execution details for the UDF from the user, FasTensor provides an internal function namely *ReportCost* (see Sect. 3.3.14 for details) to measure the time for each major part.

Fig. 3.5 The time (micro second) of running the convolution with the operator() and the *ReadNeighbors* function on the "tutorial.h5" file. Obviously, the *ReadNeighbors* is much faster than the operator() to access more than one neighborhood cell

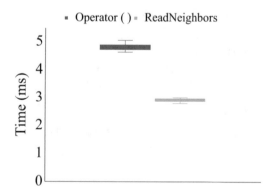

3.3 The Array Class

The **Array** class in FasTensor implements the multidimensional data model and its associated data management tasks, such as chunking. The **Array** class also implements the execution engine for the user-defined function on the multidimensional data model. The user-defined function uses the **Stencil** class in the previous section to define various data analysis operations. In this section, we introduce the implementation for the **Array** class in FasTensor. Being the same as the previous section, we only report the user-accessible API from the **Array** class in FasTensor.

Listing 3.12 gives a high-level summary of the **Array** implementation in FasTensor. The **Array** class contains a few constructors for user to create objects with different initialization information. One of the most important methods in **Array** is the *Transform* function, which runs a user-defined function on the **Array** object. It also includes other methods like *SetStride* and *SetOverlapPadding* to control the behavior of the *Transform* during the execution of user-defined function. The *AppendAttribute* function is provided by FasTensor to merge multiple arrays together for analysis. The *ControlEndpoint* function passes command to the endpoint to control its behaviors. The *ReadArray* function and the *WriteArray* function can help to access data from the object. The *Backup* function dumps data cached in memory endpoint to a file based endpoint, and the *Restore* function does the vice versa. A help function *ReportCost* is defined to report the time for internal steps such as reading the data and executing the *Transform*. Other methods are also defined to set tag (or called attributes) or to create necessary files that can be used

Listing 3.12 Overview of the **Array** class in FasTensor

```
1   template <class T>
2   class Array{
3   public:
4     //Constructors cs: chunk size, os: overlap size
5     Array();
6     Array(string endpoint_id);
7     Array(string endpoint_id, vector<int> cs);
8     Array(string endpoint_id, vector<int> cs, vector<int> os);
9     //Methods for chunk size
10    int SetChunkSize(vector<int> cs);
11    int SetChunkSizeByMem(size_t max_mem_size);
12    int SetChunkSizeByDim(int  dim_rank);
13    int GetChunkSize(vector<int> &cs);
14    //Methods for overlap (halo, ghost zone)
15    int SetOverlapSize(vector<int> os);
16    int SetOverlapSizeByDetection();
17    int GetOverlapSize(vector<int> &os);
18    int SetOverlapPadding(T padding_value);
19    int SyncOverlap();
```

Listing 3.12 (continued)

```
1            //Set, get and control endpoint
2       int SetEndpoint(string endpoint_id);
3       int GetEndpoint(string &endpoint_id);
4       int ControlEndpoint(int cmd, vector<string> &argv);
5       //Methods for attributes
6       template <class AType>
7       int AppendAttribute(string endpoint_id);
8       template <class AType>
9       int InsertAttribute(string endpoint_id, int index);
10      int GetAttribute(int index, string &endpoint_id));
11      int EraseAttribute(int index);
12      //Turn on stride during Transform, ss: stride size
13      int SetStride(vector<int> ss);
14      int GetStride(vector<int> &ss);
15      //Transform method
16      template <class UType, class BType = UType>
17      int Transform(Stencil<UType> (*UDF_P)(const Stencil<T> ),
                ↪Array<BType> *B);
18      //Read and write a subset
19      int ReadArray(vector<unsigned long long> start,
                ↪vector<unsigned long long> end, vector<T> &data);
20      int WriteArray(vector<unsigned long long> start,
                ↪vector<unsigned long long> end, vector<T> data);
21      //Dump in-memory data to an endpoint (file), or vice versa
22      int Backup(string endpoint_id);
23      int Restore(string data_endpoint_p);
24      //Methods for tag (or called property)
25      template <class PType>
26      int SetTag(string name, PType value);
27      template <class PType>
28      int GetTag(string name, PType &value);
29      int GetAllTagName(std::vector<string> &tag_names)
30      //Get/set array size
31      int GetArraySize(vector<unsigned long long> &array_size);
32      int SetArraySize(vector<unsigned long long> array_size);
33      int GetArrayRank(int &array_rank); //Get data rank
34      int CreateVisFile(FTVisType type); //Create vis. file
35      void ReportCost();//Help function to report and get costs
36      int  GetReadCost(vector<double> &cost_stats);
37      int  GetWriteCost(vector<double> &cost_stats);
38      int  GetComputingCost(vector<double> &cost_stats);
39   };
```

to create plots for the data. Details of these methods are presented in the following part of this section.

3.3.1 *Constructors of Array*

Signature:

```
Array();
Array(string endpoint_id);
Array(string endpoint_id, vector<int> cs);
Array(string endpoint_id, vector<int> cs, vector<int> os);
```

Description:

  These constructors create a new **Array** object that can be used as either an input or an output for the data analysis in FasTensor. An object without any information can be built by using the first constructor. The empty object can be assigned values later with other methods, e.g., *SetEndpoint*. The second constructor accepts a single parameter which is the endpoint identification. Except the endpoint identification, users can provide the chunk size (*cs*) as another parameter for the third constructor. In this case, the overlap size (os) will be initialized to be zero but users can change it later with the method *SetOverlapSize*. The constructor can accept both the chunk size (***cs***) and the overlapping size (***os***) as input parameters.

Parameter:

- *endpoint_id* contains the identification for the endpoint. The endpoint in FasTensor can be a file containing either the input data or the output data. The *endpoint_id* is represented by a string with the blow formation: "Endpoint type" and "Endpoint information". The "Endpoint type" in FasTensor now supports EP_HDF5, EP_PNETCDF, EP_ADIOS, EP_MEMORY, EP_BINARY, and EP_DIR. The "Endpoint information" is also a string which contains the endpoint specific information. It can also be a colon separated string, as shown in Table 3.1. A formal description of the endpoint is highlighted in the below frame.

Endpoint
In FasTensor, the Endpoint represents the data to build **Array** object for analysis. The data can be used either as input or output. The Endpoint identification contains the Endpoint type and the Endpoint specific information, which are separated by the colon ":". Usually, the Endpoint information contains the file name (as well as path) and its internal data set (or variable) identification.

- *cs* is the chunk size that is used to split the **Array** object for out-of-core or parallel processing. The *cs* is a vector where the *cs[i]* contains the chunk size for the *i*th dimension of the **Array** object. Taking a 16 by 12 **Array** as an example, it is split into 12 chunks when *cs[0] = 4* and *cs[1] =4*. In this case, these 12 chunks can be processed in either parallel or out-of-core manner, or both. An illustration of the chunk size is presented in Fig. 3.6 in following subsections.

Table 3.1 Endpoint type and specific Endpoint information in FasTensor

Endpoint type	Endpoint information
EP_HDF5	HDF5 File Name : HDF5 Dataset Name[a]
EP_PNETCDF	PNetCDF File Name: PNetCDF Variable Name
EP_ADIOS	ADIOS File Name : ADIOS Variable Name
EP_MEMORY	Not needed
EP_BINARY	Binary File Name
EP_DIR	Endpoint Type : Endpoint Information

[a]The "HDF5 Dataset Name" includes HDF5 group path

- *os* is the overlap size (or called ghost zoo size or halo size) which is used to extend each chunk during the split. Like the chunk size, the overlap size is also a vector where the *os[i]* contains the overlap size for the *i*th dimension. For a chunk that has the size *cs[i]*, FasTensor expands the size of each chunk to be *cs[i]*+ 2 × *os[i]*) during the split. Actually, the *os[i]* specify the overlap size that goes to both directions (upper and lower). Using the same example for *os* parameter above, the size for a chunk (without boundary cells) become 6 by 6 when *os[0] = 1* and *os[1] = 1*. The size for the chunk that contains boundary cells will be adjusted properly. An illustration of the overlap size is presented in Fig. 3.6 in following subsections.

Example:

Listing 3.13 gives an simple example to use the **Array** constructor to create object which points to a 2D data set "dat" within the HDF5 file "./tutorial.h5". The chunk size is (4, 4) and the overlap size is (1, 1). In this example, the fourth constructor is called to initialize the **Array** object. Users can update the endpoint ID, the chunk size and the overlap size later with their corresponding methods.

Listing 3.13 Example of using the Constructors of the **Array**

```
std::vector<int> chunk_size = {4, 4};
std::vector<int> overlap_size = {1, 1};
FT::Array<float> *A = new
    ↪FT::Array<float>("EP_HDF5:./tutorial.h5:/dat",
    ↪chunk_size, overlap_size);
```

3.3.2 SetChunkSize, SetChunkSizeByMem, SetChunkSizeByDim, and GetChunkSize

```
int SetChunkSize(vector<int> cs);
int SetChunkSizeByMem(size_t max_mem_size);
int SetChunkSizeByDim(int  dim_rank);
int GetChunkSize(vector<int> &cs);
```

Description:

These methods provide different ways for users to set and get the chunk size of an **Array** object. All these methods are lazy-evaluation functions which will be in effect until the call the *Transform* function on the **Array** object. In other words, the chunk size set by these methods are used by the *Transform* function at the beginning of its run. When multiple methods are called, the last call will overwrite previous ones and thus the chunk size from the last one will be used. The *SetChunkSizeByMem* function uses the automatic chunking method by FasTensor which takes the **max_mem_size** as one limit of the factor. The details of the automatic chunking method is presented in the work [8]. The general idea of the method is to minimize the I/O cost via reading the data along their row-major order as much as possible. The *SetChunkSizeByDim* function asks FasTensor to split the array on a particular dimension evenly on all parallel processes. Because of the lazy evaluation, the *GetChunkSize* function may only be called after the *Transform* method to obtain the correct chunk size used during the evaluation. The meaning of the chunk size and the usage of these chunking methods are demonstrated in Fig. 3.6.

Parameter:

- *cs* is a vector with the size n and the *cs[i]* contains the chunk size for the ith dimension of a n-dimensional array, where $i \in [0, n-1]$. During the *SetChunkSize* function, the **cs** vector contains the value to be assigned to an **Array** object. During the *GetChunkSize* function, the **cs** contains the value the **Array** object currently has.
- *max_mem_size* from the *SetChunkSizeByMem* function is used to specify the maximum memory size that is allowed to be used during the automatic chunking. Note that, calling the method by default enables the automatic chunking by FasTensor. The details of the algorithm is presented in the work [8]. By default, the algorithm uses the row-major order to chunk the data. Conceptually, the automatic chunking algorithm tries to maximize the contiguous disk reads by splitting lower dimensions firsts and keeping the higher dimension as contiguous as much as possible. For example, there is a 16 by 12 2D array and 2 MPI processes to run the analysis on it. In this case, the chunk size from the automatic chunking algorithm by FasTensor is the 8 by 12. Thus, each read accesses contiguous 8 rows one time for processing in the row-major ordering layout.
- *dim_rank* from the *SetChunkSizeByDim* is used to specify the dimension index to be split evenly on all parallel processes. In general, let's assume the size for an array is vector s with the size n and there are p parallel processes, the chunk size is set to be $(s[0], \ldots, s[dim\_rank]/p, \ldots, s[n-1])$. For example, there is a 16 by 12 2D array and *dim_rank = 1*, the chunk size will be set to be 16 by 6 if the number of the available MPI processes is 2.

Example:

Figure 3.14 presents one example to show how to use these chunking methods. Users can first specify a chunk size during the creation of an **Array** object. Then, it adjusts the chunking size with the SetChunkSizeByDim() method. Once the *Transform* function is called, FasTensor figures out the right chunking size to use.

Listing 3.14 Example of using the *SetChunkSize** methods on a 16 by 12 2D **Array**. Let's also assume that we run the code on 2 parallel processes. Since the overlap parameter is missing in this constructor, the overlap size will be assumed to be zero here

```
std::vector<int> chunk_size = {4, 4};
FT::Array<float> *A = new
    ↪FT::Array<float>("EP_HDF5:./tutorial.h5:/dat",
    ↪chunk_size);
A->SetChunkSizeByDim(0); //Change the chunk size to be [8, 12]
A->Transform(...);  //Lazy evaluation of the SetChunkSizeByDim
std::vector<int> chunk_size_by_dim;
A->GetChunkSize(chunk_size_by_dim); //Get the chunk size
```

3.3.3   SetOverlapSize, SetOverlapSizeByDetection, GetOverlapSize, SetOverlapPadding, and SyncOverlap

```
int SetOverlapSize(vector<int> os);
int SetOverlapSizeByDetection();
int SetOverlapPadding(T padding_value);
int GetOverlapSize(vector<int> &os);
int SyncOverlap();
```

Description:

The family of overlap methods is provided to control the parameter and behavior related to the overlap. The overlap is also called ghost zone or halo layer in other books or research papers. Users can provide the overlap size by their own to the **Array** object through calling the *SetOverlapSize* method. Users can also call the *SetOverlapSizeByDetection* to let FasTensor detect the proper overlap size to use from the user-defined function. The idea of the *SetOverlapSizeByDetection* comes from trail run method which is presented by the work [8]. Basically, FasTensor creates a special **Stencil** object as the input of the user-defined function to collect the overlap size from the user-defined code. Note that theses set functions are also lazy evaluation ones like *SetChunkSize*. So, they become in effect until the call of the *Transform* function. In terms of cells out of the boundary, the *SetOverlapPadding* function is used to assign the default values for them when the *Transform* function executes out of the boundary. Without calling the *SetOverlapPadding*, FasTensor duplicates the point right on the boundary for the one out of the boundary. The *SyncOverlap* method is called to synchronize overlap among different chunks. This processes is also called ghost zone or halo exchange. The *SyncOverlap* method only

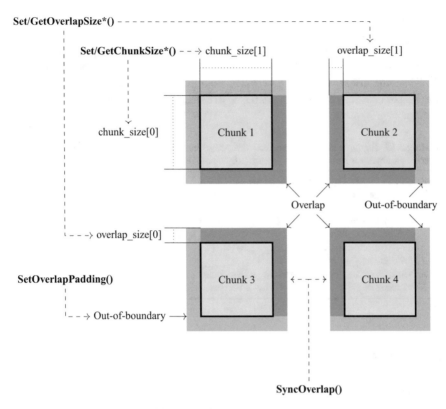

Fig. 3.6 The chunking and overlap (and their associated functions) in FasTensor. This plot uses four chunks as example. These four chunks are numbered as Chunk 1, Chunk 2, Chunk 3 and Chunk 4. Each chunk has an overlap region and out-of-boundary region. The **Set/GetChunk-Size*()** methods controls the chunk size. The **Set/GetOverlapSize*()** methods control the overlap size. The **SetOverlapPadding()** method sets the default values for out of boundary cells. The **SyncOverlap()** method exchanges overlap among chunks

be useful when the underlying endpoint is the EP_MEMORY. The usage of these overlap methods is demonstrated in Fig. 3.6.

Parameter:

- *os* is a vector with length *n*. It contains the overlap size for an **Array** object. During the call *SetOverlapSize*, it contains the overlap size to set. During the call to *GetOverlapSize*, it contains the overlap size that is currently contained by the **Array** object.
- *padding_value* sets the value to be used by FasTensor to fill out of boundary points. The *padding_value* should have the same type as the **Array** object, which is specified by the template type *T*.

Example:

One example about how to use these overlap methods is presented in Fig. 3.15. User can ignore the overlap size during the creation of the **Array** object. Then, it assigns the overlap size 1, 1 with the call of *SetOverlapSize* later. The overlap size 1, 1 become effect when the *Transform* function is called. The effect means that FasTensor uses it as the final overlap size to extend each chunk data with an overlap region.

Listing 3.15 Example of using these *SetOverlapSize** methods on a 16 by 16 2D **Array**. Since the overlap parameter is missing in this constructor, the overlap size is equal to zero at the beginning

```
std::vector<int> chunk_size = {4, 4};
//The overlap size to be [0, 0] at the beginning
FT::Array<float> *A = new
    ↪FT::Array<float>("EP_HDF5:./tutorial.h5:/dat",
    ↪chunk_size);
std::vector<int> overlap_size = {1, 1};
//Set overlap size to be [1, 1]
A->SetOverlapSize(overlap_size_new);
A->Transform(...);   //Lazy evaluation of the SetOverlapSize
```

3.3.4 Transform

Signature:

```
template <class UType, class BType = UType>
void Transform(Stencil<UType> (*UDF)(const &Stencil<T>),
               Array<BType> *B);
void Transform(Stencil<UType> (*UDF)(const &Stencil<T>),
               Array<BType> &B);
void Transform(Stencil<UType> (*UDF)(const &Stencil<T>))
```

Description:

The *Transform* method runs an user-defined function (UDF) on the **Array** object and stores the result in another **Array** object *B*. The user-defined function contains the rules to transform the current **Array** object to another **Array** object. The *UDF* has one **Stencil** object as the input and another **Stencil** object as the output. For the input **Stencil** object, it has the same element type *T* as the input **Array** object. For the output object **Stencil** object, it can has different element type *UType* from both the input and output **Array** objects. The data element type for the output **Array** object *B* is the same as that of the output **Stencil** object *UType* in most cases. But the data element type for the **Array** object *B* can also be different from the *UType*. Also, users can choose to pass the object (B) as a pointer to object or an reference to object. Note that users don't need to perform explicit instantiation on these template parameters. The C++ compiler can automatically deduce the type for the template argument and instantiate the *Transform* function.

Parameter:

- *UDF* is the function name (or called pointer in some textbook) for the user-defined function. The function pointed by the *UDF* must have the below signature:

```
Stencil<UType> (*UDF)(const &Stencil<T>)
```

where the **Stencil** class in FasTensor is an abstraction for an array cell and its neighborhood. Details of the **Stencil** class are presented in the previous Sect. 3.2. The user-defined function accepts one Stencil object as the input and another Stencil object as the output. These two **Stencil** objects can have different data element type. The template parameter *T* is the element type for the input **Stencil** object which must be the same as the one for the calling **Array** object. The template parameter *UType* is the element type for the output **Stencil** object. It is also worth noting that the input **Stencil** object is passed as an *const* object without allowing any modification on it.
- *B* is the output **Array** object to store the result from executing the *UDF* function. A *nullptr* can be assigned to *B* to drop the result but only to execute the *UDF* on the **Array**. In this case, users may declare a global variable or an **Array** object to store the results from executing the UDF function.
- *T* is the template type for the data element in the calling **Array** object itself. Any C++ basic data type like *int* or *float* can be the value of *T*. More details about the type can be found in the later Sect. 3.4.3.
- *UType* is the template type for the data element in the output **Stencil** object of the *UDF* function. Most C++ basic data types like *int* or *float* can be the value of *UType*.
- *BType* is the template type for the data element in the output **Array** B. Like the *UType*, the *BType* accepts most C++ basic data type like *int* or *float*. Note that *BType* is the same as *UType* by default. It means that type of data element in *B* is the same as the type for the data element in the output **Stencil** object (of the UDF). In some cases, *BType* may be different from *UType*. One particular example is that *UType* is a vector (i.e., *std::vector<float>*) but *BType* is a scaler type, such as *float*. In this case, FasTensor will convert the vector to be *float* typed data.

Example:

Listing 3.16 gives an example to demonstrate how to use the *Transform* method to convert a float array into an integer array. To help users understand the template parameters in *Transform*, we highlight the template type and also denote their relationship in the code too. Basically, the *UDF* parameter for *Transform* has the value *udf_convert*. The *udf_convert* parameter is the name of a function (i.e., pointer), which takes an input **Stencil** with the *int* type and outputs another **Stencil** object with the *float* type. The input array *A* has the same *int* type as the input **Stencil** object. The output array *B* has the same *float* type as the output **Stencil** object.

Listing 3.16 An example for using the *Transform* function to convert a float array to an integer array. It highlights *UType*, *BType* and *T* (as well as their relationship) in the source code

```
1
2          UType      UDF                              T
3
4            ↓         ↓                              ↓
5    Stencil<int> udf_convert(const Stencil<float> &iStencil)
6    {
7        return Stencil<int> (static_cast<int>(iStencil(0,0));
8    }
9    int main(int argc, char *argv[])
10   {
11       FT::FT_Init(argc, argv);
12       std::vector<int> cs = {4, 4}; //chunk size
13                       T
14
15                       ↓
16       FT::Array<float> A("EP_HDF5:./tutorial.h5:/dat", cs);
17           BType (= UType )
18
19                   ↓
20       FT::Array<int> B("EP_HDF5:./tutorial_convert.h5:/dat");
21       A.Transform(udf_convert, B);
22
23       FT::FT_Finalize();
24       return 0;
25   }
```

3.3.5 SetStride and GetStride

Signature:

```
int SetStride(vector<int> ss);
int GetStride(vector<int> &ss);
```

Description:

FasTensor uses the *SetStride* method to execute the user-defined function (UDF) on a few particular cells that have fixed distance from each other. The distance is specified with the stride size (*ss*) parameter. Users can also use the *GetGlobalIndex* function (in previous Sect. 3.2.8) to have more specific control on where to tun the UDF. The difference between the *SetStride* method and the *GetGlobalIndex* method is that FasTensor skips the calling the UDF in *SetStride* but FasTensor needs to call UDF in *GetGlobalIndex*. There might be some minor performance difference.

Listing 3.17 Example of applying the *SetStride* to calculate statistics per row of a 2D array

```
Stencil<std::vector<float>> udf_simple_stats(const
    ↪Stencil<float> &iStencil){
  std::vector<float> stats(3);
  stats[0] = iStencil(0, 0); //Minimum
  stats[1] = iStencil(0, 0); //Maximum
  stats[2] = 0;              //Summary
  for(int i = 0; i < 12; i++){
    stats[0] = (iStencil(0, i) < stats[0])?iStencil(0,
        ↪i):stats[0];
    stats[1] = (iStencil(0, i) > stats[1])?iStencil(0,
        ↪i):stats[1];
    stats[2] += iStencil(0, i);
  }
  std::vector<size_t> shape = {1,3};
  return Stencil<std::vector<float>> (stats, shape) ;
}

int main(int argc, char *argv[]){
  FT_Init();
  Array *A =  new Array("EP_HDF5:./tutorial.h5:/dat");
  Array *B =  new Array("EP_HDF5:./tutorial_stats.h5:/dat");
  std::vector<int> ss(2) = {1, 12};
  A->SetStride(ss); //Set stride size
  A->Transform(udf_simple_stats, B);
  FT_Finalize();
}
```

Parameter:

- *ss* is a vector with the size n. The $ss[i]$, where $i \in [0, n-1]$, specifies the number of cells to skip on the ith dimension. The calculation of the stride size is zero based which means it starts from the very first cell of the chunk. The *ss* vector contains the current stride size during the call to the *GetStride* function.

Example:

An example to use the *SetStride* method is presented in Listing 3.17. This example calculates a few statistical information for each row of a 2D array. The 2D array has the size of 16 by 12. Each row can be a time series data. The *SetStride* accepts a parameter of 1 by 12. It indicates that FasTensor executes on every 12 cells on the second dimension. Basically, FasTensor executes the *udf_simple_stats* function on the first cell of each row. Hence, within the *UDF_Simple_Stats* function, it access all cells from s(0, 0) to s(0, 11). When executing the code, FasTensor executes the *udf_simple_stats* function per row and gets the minimum, maximum and summary. These results are stored into another **Array** object B. Hence, the object B has the size 16 by 3 after finishing the execution of the *Transform* function.

3.3.6   AppendAttribute, InsertAttribute, GetAttribute and EraseAttribute

Signature:

```
template <class AType>
int AppendAttribute(string endpoint_id);
template <class AType>
int InsertAttribute(string endpoint_id, int index);
int GetAttribute(int index, string &endpoint_id));
int EraseAttribute(int index);
```

Description:

Each array in FasTensor can have multiple attributes. Each attribute can be an independent multidimensional array, which is represented by an *endpoint_id*. All attributes of a single array must have the same data element type, size and rank. All these attributes have the same chunk size and overlap size. The *AppendAttribute* method appends an endpoint as the attribute of the array. The *InsertAttribute* function inserts an attribute at the *index*th position. The *EraseAttribute* function removes the *index*th attribute. The *GetAttribute* function gets the endpoint identification (id) at the *index*th position.

Parameter:

- *endpoint_id* contains the identification information of the endpoint of an attribute. Basically, it contains a colon separated string where the first part is the type of endpoint and the second part is the specific information for the endpoint. Details of the endpoint identification are presented in the previous Sect. 3.3.1.
- *AType* is a template type parameter which specifies the data element type in the endpoint. Each endpoint can have its own element type. The *AType* is used by the internal functions of FasTensor to initialize the endpoint object.
- *index* contains the zero-based position to get, insert or erase the attribute.

Example:

Listing 3.18 gives an example which uses the *AppendAttribute* method. This example has an input array A which contains three attributes: *x*, *y* and *z*. The output Array B has fours attributes, *x*, *y*, *z* and *ave*. The attribute *ave* is equal to $\frac{x+y+z}{3.0}$. FasTensor uses the C++ keyword **struct** to declare the compound data structure for all attributes of an array. This compound data structure is also used by the UDF to express the user-defined function. For the input array A, its **struct** is defined from Line 1 to Line 5. The **struct** data structure for the array B is defined from Line 6 to Line 11. Within both **struct** data structures, it contains a macro namely *FT_UDT_INIT*. The *FT_UDT_INIT* is defined by FasTensor to reflect attributes of the structure. The purpose of this reflection on the structure is to allow FasTensor to handle attributes data internally. These attributes for both A and B are stored in different HDF5 datasets, which may come from a single file or multiple files. During the user-defined function *udf_vds*, it uses both the "iStruct" and "oStruct" as the data type. Hence, the return value of iStencil(0, 0) has the type of "oStruct" from Line 16 to Line 18. The result value records *x*, *y*, and *z* from iStencil and also calculates the average from *x*, *y*, and *z*.

3.3.7   SetEndpoint and GetEndpoint

```
int SetEndpoint(string endpoint_id);
int GetEndpoint(string &endpoint_id);
```

Description:
   When an user creates an **Array** object without any constructor parameter. The *SetEndpoint* can be used to set the ***endpoint_id*** parameter. The *GetEndpoint* can be used to obtain the ***endpoint_id*** parameter.

Parameter:

* ***endpoint_id*** has the same semantic as the one used in the constructor of the **Array** in the previous Sect. 3.3.1.

3.3.8   ControlEndpoint

Signature:

```
int ControlEndpoint(int cmd, vector<string> &argv);
int ControlEndpoint(int cmd);
```

Description:
   The ***ControlEndpoint*** method passes a command (***cmd***) and its argument vector (***argv***) to the endpoint of an array object. This controls how the endpoint act when it performs data reading, writing, or other operations. Both the ***cmd*** parameter and the ***argv*** parameter are endpoint specific. For example, in the HDF5 endpoint, one typical example is to enable or disable collective I/O. A version of ***ControlEndpoint*** without the ***argv*** argument is also provided by FasTensor. A list of predefined command for the endpoint in FasTensor can be found in the following Sects. 3.3.15 and 3.3.16.

Parameter:

* ***cmd*** is an integer which represents the command passed to the endpoint. It is defined specifically for each endpoint.
* ***argv*** is a vector of string that contains arguments for a specific command. The ***argv*** can also contain certain return value by executing the ***cmd***.

Listing 3.18 Example of using *AppendAttribute* method

```
1   struct iStruct
2   {
3       FT_UDT_INIT(iStruct)
4       int x, y, z;
5   };
6   struct oStruct
7   {
8       FT_UDT_INIT(oStruct)
9       int z, y, z;
10      float ave;
11  };
12  //duplicate original data and calculate average
13  inline Stencil<oStruct> udf_vds(const Stencil<iStruct>
          ↪&iStencil)
14  {
15      oStruct ovds;
16      ovds.x = iStencil(0, 0).x;
17      ovds.y = iStencil(0, 0).y;
18      ovds.z = iStencil(0, 0).z;
19      ovds.ave = (ovds.x + ovds.y + ovds.z) / 3.0;
20      return Stencil<oStruct>(ovds);
21  }
22  int main(int argc, char *argv[])
23  {
24      FT_Init(argc, argv);
25      //The chunk size and the overlap size
26      std::vector<int> chunk_size = {4, 4};
27      //Input data
28      Array<iStruct> *A = new Array<iStruct>(chunk_size);
29      A->AppendAttribute<int>("EP_HDF5:./tutorial.h5:/x");
30      A->AppendAttribute<int>("EP_HDF5:./tutorial.h5:/y");
31      A->AppendAttribute<int>("EP_HDF5:./tutorial.h5:/z");
32      //Result data
33      Array<oStruct> *B = new Array<oStruct>();
34      B->AppendAttribute<int>("EP_HDF5:./tutorial.h5:/x");
35      B->AppendAttribute<int>("EP_HDF5:./tutorial.h5:/y");
36      B->AppendAttribute<int>("EP_HDF5:./tutorial.h5:/z");
37      B->AppendAttribute<float>("EP_HDF5:./tutorial.h5:/ave");
38      //Run
39      A->Transform(udf_vds, B);
40      //Clear
41      delete A;
42      delete B;
43      FT_Finalize();
44      return 0;
45  }
```

Example:

The code in Listing 3.19 shows how to use the **ControlEndpoint** method to choose the I/O strategy used by the HDF5 endpoint. Specifically, it can choose to use MPI I/O and Collective I/O[9] during the endpoint's I/O operations, like opening a file and reading a file. The Collection I/O has the dependency on the MPI I/O. These I/O strategies may impact the overall I/O performance.

Listing 3.19 An example shows how to use the *ControlEndpoint* method to choose I/O strategy

```
Array<float> *A =  new
    ↪Array<float>("EP_HDF5:./tutorial.h5:/dat");

//Disable collective I/O and MPI IO
A->ControlEndpoint(HDF5_DISABLE_MPI_IO);
A->ControlEndpoint(HDF5_DISABLE_COLLECTIVE_IO);

//Enable collective I/O and MPI IO
A->ControlEndpoint(HDF5_ENABLE_MPI_IO);
A->ControlEndpoint(HDF5_ENABLE_COLLECTIVE_IO);
```

3.3.9   ReadArray and WriteArray

Signature:

```
int ReadArray(vector<unsigned long long> start,
    ↪vector<unsigned long long> end, vector<T> &data);
int WriteArray(vector<unsigned long long> start,
    ↪vector<unsigned long long> end, vector<T> data);
```

Description:

The *ReadArray* function reads a contiguous array subset specified from *start* to *end*. The *WriteArray* function writes a vector of data to the array subset from *start* and *end*. In a parallel execution, each process can call both the *ReadArray* method and the *WriteArray* method with its own *start* and *end* parameters.

Parameter:

- *start* is a vector with the size *n* and the **unsigned long long** type. It specifies the starting address of the data to read or write. The rank of *start*(i.e., *start.size()*) is equal to the number of dimensions of an array. The *start* is zero based and is inclusive.
- *end* is a vector with the size *n* and the **unsigned long long** type. It specifies the ending address of the data to read or write. The rank of *end*(i.e., *end.size()*) is

---

[9]A good information source about the Collection I/O can be found here: https://www.mcs.anl.gov/projects/romio/.

equal to the number of dimensions of an array. The **end** is zero based and is inclusive.

- **data** contains the data to write or the data to read from the calling **Array** object. The **data** is flat into 1D data in the row-major ordering.

Example:

Listing 3.20 presents an example to show how to use *ReadArray* and *WriteArray* to access data in a 2D 16 by 12 array.

Listing 3.20 An example shows how to use *ReadArray* and *WriteArray* to access data from a 16 by 12 2D array in a HDF5 file

```
Array<float> *A =  new
    ↪Array<float>("EP_HDF5:./tutorial.h5:/dat");
std::vector<unsigned long long> start_addr, end_addr;
start_addr.push_back(0); //zero based and inclusive
start_addr.push_back(0); // zero based and inclusive
end_addr.push_back(11); // inclusive
end_addr.push_back(11; // inclusive

std::vector<float> data_to_write(144, 0); //All 0 values
A->WriteArray(start_addr, end_addr, data_to_write);
std::vector<float> data_to_read;
//ReadArray is responsible for the space allocation
A->ReadArray(start_addr, end_addr, data_to_read);
```

3.3.10 SetTag and GetTag

Signature:

```
template <class TType>
int SetTag(const std::string name,  TType value)
template <class TType>
int GetTag(const std::string name, TType &value)
int GetAllTagName(std::vector<string> &tag_names)
```

Description:

The Array class in FasTensor provides the *SetTag* method for users to store tag (key-value based attribute or metadata) along an array. The *GetTag* method reads **value** for a tag with **name**. The *GetAllTagName* method obtains all tag names as a vector of string. The template type **TType** represents the type of **value** for the tag.

Parameter:

- **name** is a string which contains the identification (key) for a tag.
- **value** has a template type **TType**, which contains the value for a certain tag. The **TType** can be either a scale type, such as **float** or **string**, or a vector type, such as **std::vector<float>**.

- *tag_names* is a vector of the string. It contains all tag names (i.e., keys) for an **Array** object during the call of the *GetAllTagName*.

Example:
Listing 3.21 gives an example to attach tags to an **Array** object.

Listing 3.21 An example shows how to use the *SetTag* method and the *GetTag* method to store and get attributes for a 16 by 12 2D array in HDF5 file

```
Array<float> *A =  new
    ↪Array<float>("EP_HDF5:./tutorial.h5:/dat", ...);
//Attach tags to the A
std::string user_id_key  = "user_id";
std::string user_id_value = "ft";
A->SetTag(user_id_key, user_id_value);
std::string data_size_key  = "data_size";
std::vector<int> data_size_value = {16, 12};
A->SetTag(data_size_key, data_size_value);
std::string ave_key  = "ave";
double ave_value = 10.999;
A->SetTag(ave_key, ave_value);
...
//Get them back later
std::string user_id_value_get;
A->GetTag(user_id_key, user_id_value_get);
std::vector<int> data_size_value_get;
A->GetTag(data_size_key, data_size_value_get);
double ave_value_get;
A->SetTag(ave_key, ave_value_get);
```

3.3.11   GetArraySize and SetArraySize

Signature:

```
int GetArraySize(vector<unsigned long long> &array_size);
int SetArraySize(vector<unsigned long long> array_size);
```

Description:
The *GetArraySize* method and the *SetArraySize* method are used to get and set the size of the array, respectively. *SetArraySize* is only reserved for further usage because FasTensor mostly gets the size from the endpoint (as input) or automatically infers the array size from all arguments of the *Transform* function.

Parameter:

- *array_size* is a vector with length n. It contains the size of an **Array** object.

Example:
A example in Listing 3.22 shows how to use the *SetArraySize* method to get the size of an array.

Listing 3.22 An example shows how to use the *GetArraySize* method to find the size for a 16 by 12 2D array in HDF5 file

```
Array<float> *A =  new
    ↪Array<float>("EP_HDF5:./tutorial.h5:/dat");
std::vector<unsigned long long> data_size;
A->GetArraySize(data_size);
```

3.3.12 Backup and Restore

Signature:

```
int Backup(string endpoint_id);
int Restore(string data_endpoint_p);
```

Description:

FasTensor has an in-memory endpoint which can be specified with the *EP_MEMORY* type code. In the *EP_MEMORY* endpoint, the data is cached in the memory across all available nodes. To facilitate the data migration into or out of the in-memory endpoint, FasTensor provides two functions, *Backup* and *Restore*. The *Backup* loads the data from the in-memory endpoint to another endpoint, which is mostly a file-based endpoint, such as *EP_HDF5*. The *Restore* function performs the inverse task to migrate data from a file-based endpoint to an in-memory endpoint. Both the *Backup* method and the *Restore* method are at the stage of experiment and may be tuned in the future. The in-memory endpoint may use the third-party software, such as DASH,[10] to distribute the data among MPI processes. Also, FasTensor has its own implementation for distributed array data structure.

Parameter:

- *endpoint_id* is a string which contains the identification of an endpoint. Please refer to the previous Sect. 3.3.1 for the semantic of the endpoint identification.

Example:

Listing 3.23 demonstrates the usage of the *Backup* method. This example reads data from a HDF5 file and then applies two user-defined functions on the data via the *Transform* function. The intermediate Array is stored in memory. Meanwhile, the intermediate array is dumped into a HDF5 file endpoint as check point through the *Backup* method. Finally, the result is stored into another HDF5 file on disk for preservation. The example accepts two user-defined functions, *udf_cache1* and *udf_cache2*. Both functions multiply each point with 2.0. The *udf_cache1* function is applied onto the Array A. The output data from the *udf_cache1* function is stored

---

[10]The DASH is a C++ Template Library for distributed array data structures with the support for HPC and data-driven Science. More details can be found here: https://github.com/dash-project.

Listing 3.23   An example of cache data in memory in FasTensor

```
1   inline Stencil<float> udf_cache1(const Stencil<float> &iStencil)
2   {
3       return Stencil<float> (iStencil(0, 0) * 2.0);
4   }
5   inline Stencil<float> udf_cache2(const Stencil<float> &iStencil)
6   {
7       return Stencil<float> (iStencil(0, 0) * 2.0);
8   }
9   int main(int argc, char *argv[])
10  {
11      FT_Init(argc, argv);
12      std::vector<int> cs = {4, 12}; //chunk size
13      std::vector<int> os = {0, 0}; //overlap size
14      Array<float> *A = new
            ↪Array<float>("EP_HDF5:./tutorial.h5:/dat", cs, os);
15      Array<float> *B = new Array<float>("EP_MEMORY");
16      Array<float> *C = new
            ↪Array<float>("EP_HDF5:./tutorial_final.h5:/dat");
17      A->Transform(udf_cache1, B);
18      //Backup intermediate data
19      B->Backup("EP_HDF5:./tutorial_cache_inter.h5:/dat");
20      B->Transform(udf_cache2, C);
21      delete A;
22      delete B;
23      delete C;
24      FT_Finalize();
25      return 0;
26  }
```

into the Array B. The Array B is initialized with the "EP_MEMORY" in line 15. Since FasTensor internally deals with the chunk and overlap size, there is no need to have them as initialization parameters for the Array B. The Array C (in line 16) is initialized with a HDF5 file based endpoint to contain the final array data. In line 17, the *udf_cache1* function is applied onto the Array A. Since the output of this *Transform* is array B, which is in memory, the output the *udf_cache1* is stored in memory. The Array B is dumped into a HDF5 endpoint in line 19.

3.3.13   CreateVisFile

Signature:

```
int CreateVisFile(FTVisType type);
int CreateVisFile();
```

Description:

The *CreateVisFile* method from FasTensor creates necessary files to produce visualization for the Array object. It is worth noting that FasTensor does not support picture render and plot but it provides necessary files to help users to generate nice result figures.

Parameter:

- *type* has the type of *FTVisType*. In FasTensor, it currently has the values, *FT_VIS_XDMF*, *EP_VIS_PYTHON* and *EP_VIS_R*. For the one without the *type* argument, it uses the *FT_VIS_XDMF* by default. The *FT_VIS_XDMF* produces the XDMF[11] file that is widely used to visualize scientific data on a HPC system. The XDMF file can be directly opened by software, such as ParaView[12] and VisIt,[13] to produce multidimensional images.

Example:

The example of using the *CreateVisFile* function is presented in Listing 3.24. After executing the code, two files, "tutorial.h5.xdmf" and "tutorial_tran.h5.xdmf", are created on the side of the "tutorial.h5" and "tutorial_tran.h5" by appending the ".xdmf" extension. Both XDMF files can be opened by the ParaView to generate the figures. The result figures are presented in Fig. 3.7.

3.3.14   ReportCost

Signature:

```
void  ReportCost();
int   GetReadCost(vector<double> &cost_stats);
int   GetWriteCost(vector<double> &cost_stats);
int   GetComputingCost(vector<double> &cost_stats);
```

Description:

FasTensor runs the user-defined function(UDF) via its *Transform* function. It hides all execution details from users. FasTensor provides a few help functions to track the cost of these execution. *ReportCost* reports the cost of reading data, executing the UDF and writing data across all processes. Also, *GetReadCost*, *GetWriteCost*, and *GetComputingCost* returns the cost for reading data, writing data, and executing UDF, respectively. All these costs are measured with second unit.

---

[11] XDMF:https://www.xdmf.org/index.php/XDMF_Model_and_Format

[12] ParaView: https://www.paraview.org/

[13] VisIt: https://wci.llnl.gov/simulation/computer-codes/visit/

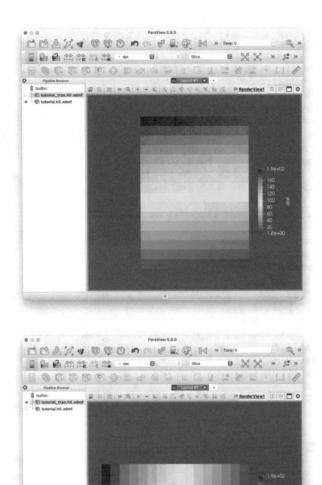

Fig. 3.7 Figures produced by ParaView using the files created by FasTensor in the Listing 3.24. The above figure is produced from the original data by opening the "tutorial.h5.xdmf" in the ParaView. The below figure is produced from the transposed data by opening the "tutorial_tran.h5.xdmf" in the ParaView

Listing 3.24 An example shows how to use the *CreateVisFile* function to plot a 16 by 12 2D array and its transposed one

```
1   inline Stencil<vector<float>> udf_tran(const Stencil<float>
        ↪ &iStencil)
2   {
3       std::vector<int> start = {0, 0}, end = {15, 11};
4       std::vector<float> data_ori, data_tra(16 * 12);
5       iStencil.ReadNeighbors(start, end, data_ori);
6       transpose_data_2D(data_ori.data(), data_tra.data(), 16, 12);
7       std::vector<size_t> shape = {12, 16};
8       return Stencil<vector<float>>(data_tra, shape);
9   }
10
11  int main(int argc, char *argv[]) // start here
12  {
13      FT_Init(argc, argv);
14      std::vector<int> chunk_size = {16, 12};
15      std::vector<int> ss = {16, 12};
16      Array<float> *A = new
            ↪Array<float>("EP_HDF5:tutorial.h5:/dat", chunk_size);
17      A->SetStride(ss);
18      Array<float> *B = new
            ↪Array<float>("EP_HDF5:tutorial_tran.h5:/dat");
19      A->Transform(udf_tran, B);
20      A->CreateVisFile();
21      B->CreateVisFile();
22      delete A;
23      delete B;
24      FT_Finalize();
25      return 0;
26  }
```

Parameter:

- *cost_stats* is a vector with the length of *three*. The *cost_stats[0]* element contains the maximum cost across all processes. The *cost_stats[1]* element contains the minimum cost across all processes. The *cost_stats[2]* element contains the average cost across all processes.

Example:

An example of using the *ReportCost* function is presented in Listing 3.25.

3.3.15 EP_DIR Endpoint

The EP_DIR endpoint in FasTensor is a special one which points to a directory. The directory may contain many files. In this section, we presents an example in

Listing 3.25 An example shows how to use the ReportCost function to get the execution overhead of the Transform function

```
1   Array<float> *A =  new
       ↪Array<float>("EP_HDF5:./tutorial.h5:/dat");
2   A->Transform(...); //Ignore the UDF function here
3   A->ReportCost();
4   //Sample output from ReportCost
5   //Read  time(s): max=0.000526, min=0.000526, ave=0.000526
6   //UDF   time(s): max=0.000104, min=0.000104, ave=0.000104
7   //Write time(s): max=0.00866, min=0.00866, ave=0.00866
```

Listing 3.26 to demonstrate how to use the EP_DIR endpoint to process lots of files when these files have the same structure and are stored within a single directory. It is worth noting that FasTensor also supports that the files under the same directory have different sizes. In this example, the same structure means that the array data stored in these files have the same number of dimensions and each dimension has the same size. Moreover, authors can apply an regular expression (regex) pattern on the both input and output file names. Specifically, for input files, the regex can be used to select proper ones as the input. For output files, the regex can be used to convert input file names to proper output file names.

In this example, we assume that a directory namely "./tutorial_dir" contains five HDF5 files to process. The names for these files are "fil1.h5", "file2.h5", "file3.h5", "file4.h5" and "file5.h5". Each file has a 16 by 12 2D array with the same data set name "*/dat*". Our example code only multiplies each value in these arrays by 2.0. However, more complex data structures can be built on top of the EP_DIR endpoint. We also apply the regex pattern to filter input files and also to specify file names for output files. Details of the example are presented in following paragraphs.

In Line 11, we show how to create the **Array** object A to represent a directory and its files. Here we have the string: "EP_DIR:EP_HDF5:./tutorial_dir:/dat" as the **endpoint_id**. The "EP_DIR" tells the **Array** object to use an endpoint that has the directory type. The "EP_HDF5" declares that the files within the directory are all HDF5 files. The "./tutorial_dir" is the directory name. The "*/dat*" is the dataset name in all these HDF5 files. An error will happen if different HDF5 files have different dataset names in this case. The **endpoint_id** for the **Array** object ignores the file name because FasTensor automatically lists all files under the directory and inserts the file name to the **endpoint_id** for the underlying HDF5_DIR endpoint. In summary, the format for the **endpoint_id** of this example is summarized in Fig. 3.8.

Recall that each array in this example has the size of 16 by 12. The chunk size at the Line 11 is set to be "16, 12", which is exactly same as the size of each 2D array. By taking this chunk size, FasTensor actually views a single file as a chunk of a large array. Then, FasTensor schedules these files (chunks) among MPI processes to execute the user-defined function *udf_dir* on them. The overlap size in this example is "0, 0" since we only apply UDF on a single point.

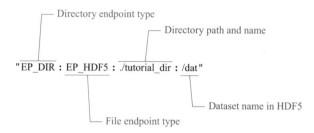

Fig. 3.8   Meaning of each part for the ***endpoint_id*** during the EP_DIR endpoint

In line 14, the **Array** object A accepts one parameter to filter its input files via an regex pattern. The syntax specifications for the regex pattern is the ECMAScript[14] which is a part of C++ standard library. In this example, it has the regex pattern `"^(.*)[135]\\.h5$"`. This pattern searches the files whose name ends with the ".h5" and also contains the "1", "3", or "5" character right before the ".h5". The command "DIR_INPUT_SEARCH_RGX" is defined by the EP_DIR endpoint during FasTensor to execute the search on input files. Internally, the "std::regex_search" from C++ library is used to perform this search function on the input file names.

In terms of the output, the **Array** object B also accepts a regex pattern to reuse the input file name but append each file name string with a "_output" string. For this purpose, the code from Line 18 to Line 20 defines the regex pattern on the output file name. When there is no regex specified for output file names, FasTensor will assume that the output file name is the same as the input file name (because they are in different directory). Internally, FasTensor uses the "std::regex_replace" from the C++ library to find all matches of the regex pattern and replace them with the new format. The first parameter in line 19 defines the regex pattern for match and the second parameter in line 20 defines the replacement format.

Finally, the user-defined function *udf_dir* is applied onto the **Array** object A via the *Transform* method. Results are stored into the **Array** object B. The example code can be compiled easily as other FasTensor programs. The compiled code can be used to run either sequentially or in parallel. Of course, we only have three files to process in this example. Hence, one can run the code with at most three MPI processes in parallel. Otherwise, computing resources are wasted.

Beyond this example, the blow list summarizes all defined commands supported by the EP_DIR endpoint in FasTensor. Each command can be passed to the EP_DIR endpoint via the *ControlEndpoint* method, as shown in the previous Sect. 3.3.8.

- ***DIR_MERGE_INDEX*** controls the index of the dimension to merge all files under a directory. The argument for the ***DIR_MERGE_INDEX*** is a integer, such as 0 and 1. The Fig. 3.9 presents an example that shows how to merge three 2D arrays in files.

---

[14]http://www.cplusplus.com/reference/regex/ECMAScript/

Listing 3.26 An example of using *EP_DIR* in FasTensor

```
1   inline Stencil<float> udf_dir(const Stencil<float> &iStencil)
2   {
3       return Stencil<float> (iStencil(0, 0) * 2.0);
4   }
5   int main(int argc, char *argv[])
6   {
7       FT_Init(argc, argv);
8       std::vector<int> cs = {16, 12};
9
10      //Input data
11      Array<float> *A = new
            ↪Array<float>("EP_DIR:EP_HDF5:./tutorial_dir:/dat",
            ↪cs);
12      std::vector<std::string> input_search_rgx;
13      input_search_rgx.push_back("^(.*)[135]\\.h5$");
14      A->ControlEndpoint(DIR_INPUT_SEARCH_RGX,
            ↪input_search_rgx);
15
16      //Output data
17      Array<float> *B = new
            ↪Array<float>("EP_DIR:EP_HDF5:./tutorial_dir_output:
            ↪/dat");
18      std::vector<std::string> aug_output_replace_arg;
19      aug_output_replace_arg.push_back("^(.*)\\.h5$");
20      aug_output_replace_arg.push_back("$1_output.h5");
21      B->ControlEndpoint(DIR_OUPUT_REPLACE_RGX,
            ↪aug_output_replace_arg);
22
23      A->Transform(udf_dir, B);
24
25      delete A;
26      delete B;
27      FT_Finalize();
28      return 0;
29  }
```

- **DIR_SUB_CMD_ARG** passes a control command and its arguments to the sub endpoint. The sub endpoint is the endpoint (i.e., file) in the directory. The first argument for the **DIR_SUB_CMD_ARG** command is a command (an integer) to the sub endpoint. The second and following arguments can be the arguments required by the command for the sub endpoint.
- **DIR_INPUT_SEARCH_RGX** passes an regex string to the EP_DIR endpoint and it also enables the regex based filter on input file names. The example of using the **DIR_INPUT_SEARCH_RGX** command is presented in Listing 3.26.
- **DIR_OUPUT_REPLACE_RGX** passes an regex string to the EP_DIR endpoint and it also enables the regex based match and replace on the output files. The first argument is the regex match pattern and the second argument is the regex replace

pattern. The example of using the ***DIR_OUPUT_REPLACE_RGX*** command is presented in Listing 3.26.

- ***DIR_FILE_SORT_INDEXES*** passes a string to an EP_DIR endpoint. The string contains the comma-separated integer, which is used to order the files populated from the EP_DIR endpoint. For example, when an argument is "1,3,2" and EP_DIR endpoint has three file (file1, file2, file3) populated from the target directory, the EP_DIR endpoint will reorder its file list to be (file1, file3, file2). When the argument has less number of indexes than that of available files, the EP_DIR endpoint will cut the file list. For example, when an argument is "1,3" and EP_DIR endpoint has three file (file1, file2, file3), the EP_DIR endpoint will reorder and cut its file list to be (file1, file3). FasTensor provides am utility function namely *Vector2String* to help convert an integer vector to a comma-separated string. The signature for the *Vector2String* is below:

```
template <typename T>
std::string Vector2String(const std::vector<T> &vec)
```

- ***DIR_INPUT_ELASTIC_SIZE*** indicates the files for the EP_DIR endpoint have varying sizes. By seeing this command, FasTensor will overlook the chunk sizes by the **Array** object and treats each input files as an independent chunk. Inside the UDF, users may use the *GetOffsetUpper* function and the *GetOffsetLower* function to obtain the size of a chunk (i.e., a file within the directory).
- ***DIR_OUTPUT_ELASTIC_SIZE*** tells FasTensor that output files have varying sizes. In current FasTensor, it can only be used when each chunk only has a single UDF instance to run (via the *SetStride* function). This UDF produces only a single output (i.e., a std::vector). This simplifies the data management inside FasTensor. This single output will be stored into a single file in the output directory. The input files for the ***DIR_OUTPUT_ELASTIC_SIZE*** command may have the same size or different sizes.

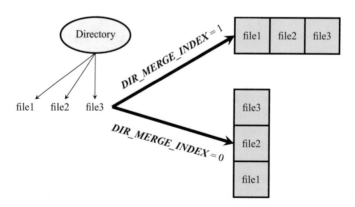

Fig. 3.9 Using the ***DIR_MERGE_INDEX*** command to merge 3 files (2D arrays) for an EP_DIR Endpoint. When ***DIR_MERGE_INDEX*** = 0, it merges files along the first dimension. When ***DIR_MERGE_INDEX*** = 1, it merges files along the second dimension

Table 3.2 List of pre-defined commands in the EP_HDF5 endpoint and other endpoints. The items marked with the "*" symbol are experimental functions in FasTensor and these items may have some changes in future. See more details about commands for EP_DIR in the previous section

Endpoint type	Pre-defined command	Argument	Functions
EP_HDF5	HDF5_ENABLE_MPI_IO	None	Enable MPI IO
	HDF5_DISABLE_MPI_IO	None	Disable MPI IO
	HDF5_ENABLE_COLLECTIVE_IO	None	Disable collective I/O
	HDF5_DISABLE_COLLECTIVE_IO	None	Enable collective I/O
EP_MEMORY	MEMORY_BACKUP	Endpoint id	Backup to another endpoint
	MEMORY_RESTORE	Endpoint id	Load into memory endpoint
	MEMORY_SYNC_OVERLAP	None	Synchronize overlap zone
	MEMORY_CLONE*	None	Create local mirrors
	MEMORY_MERGE*	None	Merge local mirrors
EP_BINARY	BINARY_SET_SIZE*	Size vector	Set size for data
	BINARY_ENABLE_TRAN_READ	None	Transpose data on read
	BINARY_DISABLE_TRAN_READ	None	Disable transpose on read
	BINARY_ENABLE_TRAN_WRITE	None	Transpose data on write
	BINARY_DISABLE_TRAN_WRITE	None	Disable transpose on write
EP_PNETCDF	None		
EP_ADIOS	None		
EP_DIR	DIR_MERGE_INDEX	Index	Order to merge
	DIR_SUB_CMD_ARG	Cmd and argv	For sub endpoint
	DIR_INPUT_SEARCH_RGX	regex pattern	Filter input
	DIR_OUPUT_REPLACE_RGX	regex pattern	Match and replace output
	DIR_FILE_SORT_INDEXES	List of index	Order input file
	DIR_INPUT_ELASTIC_SIZE	None	Enable varying input
	DIR_OUTPUT_ELASTIC_SIZE	None	Enable varying output
	DIR_OUTPUT_FILE_NAMES	List of names	Set output names

- **DIR_OUTPUT_FILE_NAMES** command gives a vector of file names to use for the output **Array** object that is built with a EP_DIR endpoint. The **DIR_UTPUT_FILE_NAMES** command needs an argument which contains the vector of file name string. Based on the scheduling method in FasTensor, it picks the file name by the order to store the results.

3.3.16   EP_HDF5 and Other Endpoints

Table 3.2 summarizes all predefined commands for endpoints in FasTensor. These commands can be used by the *ControlEndpoint* method of the **Array** class to have more detailed control on an endpoint.

3.4   Other Functions in FasTensor

3.4.1   FT_Init

Signature:

```
int FT_Init(int argc, char *argv[]);
int FT_Init(int argc, char *argv[], MPI_Comm mpi_comm);
```

Description:

FT_Init initializes the running environment for FasTensor. *FT_Init* takes two arguments from the *main* function. Internally, *FT_Init* initializes MPI environment and other necessary functions. By default, *FT_Init* uses **MPI_COMM_WORLD** to perform MPI-related communication. Users can pass a different communicator via the **mpi_comm** parameter.

Parameter:

- **argc** contains the number of element in **argv**.
- **argv** contains the pointer to arguments.
- **mpi_comm** contains the MPI communicator identification.

3.4.2   FT_Finalize

Signature:

```
int FT_Finalize();
```

Description:

FT_Finalize cleans up the runtime environment of FasTensor. One major task in *FT_Finalize* is to clean all endpoints and MPI environment. Other tasks may be added in future to the *FT_Finalize* function.

3.4.3   Data Types in FasTensor

Table 3.3 lists all current supported data types FasTensor. In most cases, users don't need to handle these types from FasTensor. But users need to assign right template type to the **Array** class and the **Stencil** class, which also need to match with the data element type in corresponding data endpoints, such as HDF5, ADIOS and PNetCDF. Internally, FasTensor converts the template type to or from the data element type in an endpoint.

Table 3.3 The data type supported in FasTensor and its endpoints, such as HDF5, ADIOS, and PNetCDF. It also lists all corresponding C++ types. These C++ types are mostly used by users to specify template type in the **Array** class and the **Stencil** class

C++	FasTensor	HDF5	ADIOS	PNetCDF
short	AU_SHORT	H5T_STD_I16LE	adios2_type_int16_t	NC_SHORT
int	AU_INT	H5T_STD_I32LE	adios2_type_int32_t	NC_INT
long	AU_LONG	H5T_STD_I64LE	adios2_type_int64_t	NC_LONG
long long	AU_LONG_LONG	H5T_STD_I64LE	adios2_type_int64_t	NC_LONG
unsigned short	AU_USHORT	H5T_STD_U16LE	adios2_type_uint16_t	NC_USHORT
unsigned int	AU_UINT	H5T_STD_U32LE	adios2_type_uint32_t	NC_UINT
unsigned long	AU_ULONG	H5T_STD_U64LE	adios2_type_uint64_t	NC_UINT64
unsigned long long	AU_ULLONG	H5T_STD_U64LE	adios2_type_uint64_t	NC_UINT64
float	AU_FLOAT	H5T_IEEE_F32LE	adios2_type_float	NC_FLOAT
double	AU_DOUBLE	H5T_IEEE_F32LE	adios2_type_double	NC_DOUBLE
std::complex<double>	AU_DOUBLE_COMPLEX	–	–	–
std::string	AU_STRING	–	–	–

Chapter 4
FasTensor in Real Scientific Applications

In this chapter, we describe two scientific applications to demonstrate the capability of FasTensor. The first application is from the earth science, where we show how FasTensor implement the self-similarity computation to detect useful signal from the data collected with distributed acoustic sensing (DAS). The second application is from plasma physics, where we apply FasTensor to analyze the hydro data from a VPIC simulation. The description in this chapter emphasis the use of FasTensor in analysis tasks, while additional information about the science is available in the published literature [5, 6].

4.1 DAS: Distributed Acoustic Sensing

DAS technology and its applications. Distributed acoustic sensing (DAS) [90] is an emerging technology that utilizes optic fiber cables to measure strain along the cable. Developed in 1990s, the idea of using optical fiber for distributed strain sensing has gained wide attention [91]. Based on the Rayleigh scatter of light, the core of the DAS technology is the optical time-domain reflectometry (OTDR). OTDR is originally developed to test and study the attenuation of optic fiber. When a laser pulse is sent along the optic fiber, the light is attenuated as it propagates. Due to Rayleigh scattering, it also produces reflected (or backscattered) light. OTDR measures the intensity of this reflection as a function of time. Because the intensity of the lights from different portions along the fiber show high divergence, analyzing the intensity of the reflected light can deduce the strain and temperature variations along the optic fiber cable.

Currently, the most mature OTDR method is the phase-sensitive coherent optical time-domain reflectometry, or Φ-OTDR for short. For distributed acoustic sensing, Φ-OTDR has high sensitivity and spatial resolution. It uses the laser light with long coherence length compared to the pulse length. By analyzing the reflected lights, it

B. Dong et al., *User-Defined Tensor Data Analysis*, SpringerBriefs in Computer Science, https://doi.org/10.1007/978-3-030-70750-7_4

is possible to find the acoustic wave from objects close to the fiber optic cable. It is able to differentiate motions of cable segments that are a meter apart. Therefore, Φ-OTDR can transform the optical fiber cable into a densely sampled sensor array. In the DAS community, these sensors are also called channels. Compared with traditional methods, like the drilling or the geophone sensor [92], DAS requires low-cost and low-maintenance, hence enables many scientific explorations that were previously infeasible.

DAS provides acoustic sensing at high resolution and large spatial extent. It detects the acoustic wave in different locations almost at the same time. These distinctive features make DAS very useful for many scientific applications. For example, Lindsey et al. [93] used DAS to catalog and analyze earthquake observations. Dou et al. [92] used DAS to monitor the earth surface to provide early warning of near-surface hazards. The earth surface refers the top tens of meters of subsurface that carries the buildings and other infrastructures, and the hazard in near-surface may damage these infrastructures. Many other applications of DAS including oil, gas, infrastructure and security, have been documented in literature [94].

Data analysis challenges. One major challenge comes with the DAS technology is how to extract intrinsic knowledge from the large volumes of data. DAS can monitor any strain along the fiber and therefore capture the vibration or wave (as well as noisy) from many objects, such as cars or earthquake, as illustrated in the left part of the Fig. 4.1. The iDAS [90] in the figure is the system to send light source, receive the reflected light, and convert the intensity of the reflected light to strain measurements stored in numerical arrays. Usually, the optical fiber can run tens of kilometers and the iDAS can send and receive light pulse at high frequency (e.g., hundreds of Hz). This rate to send and receive the light pulse is also known as the sampling rate. For example, the DAS deployment in West Sacramento, California, spans 25 kilometer (km) [90] going from Sacramento to Woodland. This installation has 2 meters spatial resolutions and 500 HZ sampling rate, which giving around 1 Terabyte of data per day. In this case, every 2 meter of the table could be regarded as a sensors or channel. The data for each minute is stored in a separate file, creating over a thousand files per day. To effectively work with the large number data files, the common solution is to develop custom analysis code on a high-performance computing system. Such a custom analysis code is time-consuming to develop, which causes the application scientist to settle for less efficient solutions, such as using matlab scripts or python or some commercial data processing framework.

There are a variety of geoscience applications that could make use of DAS data, and some of these applications require significant amounts of compute time. For instance, the local similarity may be computed to detect earthquakes, and the Fast Fourier Transform (FFT) may be applied to explore the frequency domain. In one use case, a group of geoscientist were using parallel matlab script to compute the local similarity on small subset of the DAS data but found it takes weeks of run time on an hour's worth of data. Next in this section, we will demonstrate how to use FasTensor to perform this local similarity calculation in much less time with a minimal amount of programming.

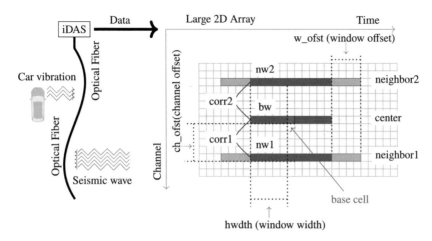

Fig. 4.1 An illustration of DAS and a data analysis example. The iDAS device monitors the strain on the fiber optic cable and produces a 2D array containing the strain measurements over time (horizontal) for each channel (vertical). The figure on the right is an example of computing similarity between the center channel and its two neighbors, *ch_ofst* apart. The computation involves a base window (bw) and two neighboring windows, nw1 and nw2. Each of these windows has $2 \times hwdth + 1$ cells. We attribute this similarity to the center of the base window and store the output of computation in a 2D array at the corresponding cell. Additional similarity computation may be performed by shifting the neighboring windows to the left or right (i.e., in time) by the amount specified by *w_ofst*

Correlation Function in FasTensor. DAS data sets consist of records of strain-rate variations along optic fiber cables caused by vibrations from sources such as vehicles, machinery, earthquakes, explosions, and so on. In geophysics, distinguishing between these sources is critical for scientific studies, such as subsurface imaging and earthquake detection. To distinguish these vibration sources, one common data analysis operation is to compute the self-similarity [95]. This self-similarity calculation breaks DAS data into subsets (from different parts of a fiber) and calculates the correlation among these subsets. When a coherent signal hits the optic fiber, the correlation among these subsets will be strong, otherwise, the correlation tend to be weak. Formally, let x and y be two signal vectors, their similarity is defined as

$$\text{Corr}(x, y) = \frac{|xy^T|}{\|x\|_2 \|y\|_2}, \tag{4.1}$$

where the $\|x\|_2$ and the $\|y\|_2$ are the root-sum-of-squares (RSS) for the vector x and y, respectively. The two vectors are similar to each other when x is close to y or $-y$. In this case, the value of the *Corr* will be close to 1 or -1. When x is significantly different from the y and $-y$, Corr is close to 0.

Figure 4.1 contains an illustration of this self-similarity calculation on DAS data shown as a dense 2D array. This calculation divides the DAS data along the time domain into windows per channel and determines the correlations between different channels [96]. Specifically, it creates a base window *bw* with $2 \times hwdth + 1$ points at the center channel. At same time, it creates two neighboring windows *nw*1 and *nw*2 on nearby channel with the offset *ch_ofst* and *-ch_ofst*, respectively. The two neighboring windows can also shift *w_ofst* on either direction in time.

The Listing 4.1 presents an implementation for the self-similarity calculation in FasTensor. To simplify the example, we have hard-coded most of the parameters involved, for example, the four parameters *hwdth*, *c_n*, *ch_ofst* and *w_ofst* at the top of the list, where *hwdth* is the offset of the channel, *c_n* is the total number of points for the window, *ch_ofst* denotes the offset of the channel, i.e., the neighboring window, and *w_ofst* is the offset for the window. The main calculation is carried out in function *udf_similarity*, which implements Eq. 4.1. This pseudo-code listing has two functions: the user-defined calculation in *udf_similarity* and the requisite function *main* to supply data and execute *udf_similarity*. The function *udf_similarity* defines calculation on each cell of the input data, and the FasTensor run-time engine automatically executes *udf_similarity* on the whole array in parallel.

Within *udf_similarity*, it reads the base window (*bw*) around the base cell. The total number of cells in the base window is $2 \times hwdth + 1$, which includes *hwdth* cells on the left, *hwdth* cells on the right as well as the base cell itself (in line 13). It calculates the root-sum-of-squares in lines 15 and 16 for the base window. The code in line 17 reads the data from the neighboring windows, *nw1* and *nw2*, at the offset *ch_ofst* and *-ch_ofst*, respectively. The two neighboring nw1 and *nw2* are moved within a prescribed range defined by the window offset *w_ofst* (see an illustration in the Fig. 4.1). This implementation reads the necessary cell values for all neighboring windows to reduce the cost of invoking *c(ch_ofst, j)* to perform I/O operations.

After obtaining the data for the neighboring windows, the function *udf_similarity* continues to calculate the correlation between the *bw* and these two neighboring channels separately. In line 22, *udf_similarity* finds the maximum correlations for each neighboring channel. Specifically, the *for* loop iterates through the neighboring windows from the left to the right. Each neighboring window has the same size *2*hwdth+1* as the base window. The variable *max_corr1* records the maximum correlation between *bw* and *nw1*; and the variable *max_corr2* records the maximum correlation between *bw* and *nw2*. After these maximum correlations are found, their average value is returned from *udf_similarity*. This average value is considered as the correlation at the base cell and is assigned to an output array at this cell location. In the following *main* function, FasTensor apply these calculation onto the whole array. Obviously, the overall self-similarity data analysis on the DAS data is extremely computing intensive.

Listing 4.1 FasTensor implementation of the self-similarity calculation. We initialize the code with the parameters from Fig. 4.1

```
1   #include "ft.h"
2   int  hwdth=5; //window width
3   int ch_ofst=5; //channel offset
4   int w_ofst=4; //window offset
5   inline Stencil<float> udf_similarity(const Stencil<short> &c)
6   {
7       int c_n=2*hwdth+1;
8       float max_corr1=0, max_corr2=0, corr1, corr2;
9       float c_sqSum, n_sqSum1, n_sqSum2, cn_Sum1, cn_Sum2;
10      std::vector<float> bw(c_n); //base window
11      std::vector<float> nw1(c_n+2*w_ofst);//neighbor window 1
12      std::vector<float> nw2(c_n+2*w_ofst);//neighbor window 2
13      for (int i=-hwdth; i<=hwdth; i++)
14          bw[i+hwdth] = c(0, i);
15      c_sqSum=inner_product(bw.begin(),bw.end(),bw.begin(),0);
16      c_sqSum=sqrt(c_sqSum);
17      for (int j=-hwdth-w_ofst; j<=hwdth+w_ofst; j++)
18      {
19          nw1[j+hwdth+w_ofst] = c(ch_ofst, j);
20          nw2[j+hwdth+w_ofst] = c(-ch_ofst, j);
21      }
22      for (int j=-w_ofst; j<=w_ofst; j++)
23      {
24          int nws = j+w_ofst;
25          cn_Sum1 = inner_product(bw.begin(), bw.end(),
                ↪nw1.begin()+nws, 0);
26          cn_Sum2 = inner_product(bw.begin(), bw.end(),
                ↪nw2.begin()+nws, 0);
27          n_sqSum1 = sqrt(inner_product(nw1.begin()+nws,
                ↪nw1.begin()+nws+c_n, nw1.begin()+nws, 0));
28          n_sqSum2 = sqrt(inner_product(nw2.begin()+nws,
                ↪nw2.begin()+nws+c_n, nw2.begin()+nws, 0));
29          corr1 = abs(cn_Sum1)/(c_sqSum*n_sqSum1);
30          corr2 = abs(cn_Sum2)/(c_sqSum*n_sqSum2);
31          if (corr1 > max_corr1) max_corr1 = corr1;
32          if (corr2 > max_corr2) max_corr2 = corr2;
33      }
34      return Stencil<float>((max_corr1+max_corr2)/2);
35  }
36  // To reduce the code size, we have hard-coded parameters
        ↪for array size 11648 by 30000 (channel by time).
37  int main(int argc, char *argv[])
38  {
39      std::vector<int> os = {5, 0};//overlap size os[0]=ch_ofst
40      std::vector<int> cs = {2912, 30000}; //chunk size
41      FT_Init(argc, argv);
42      FT::Array<short> *IFILE = new
            ↪FT::Array<short>("EP_HDF5:../das.h5:/dat", cs, os);
```

Listing 4.1 (continued)

```
1            FT::Array<float> *OFILE = new
                 ↪FT::Array<float>("EP_HDF5:./das_similarity.h5:/dat");
2            IFILE->Transform(udf_similarity, OFILE);
3            delete IFILE;
4            delete OFILE;
5            FT_Finalize();
6            return 0;
7        }
```

In the function *main*, starting in line 37, FasTensor initialize the objects for the DAS data. In this example, we assume that the data is stored in a dataset namely "dat" within a HDF5 file "das.h5". The dataset "dat" is a 2D array with the size 11,648 by 30,000, where 11,648 is the number of channels and the 30,000 is the number of sampling points in one minute. We assign the chunk size(*cs*) to be 2912, 30,000, which splits the data into four chunks based on channels. At the same time, the overlap size(*os*) is set to be 5, 0, which extends each chunk 5 rows in both directions. Since the *cs[1]* covers the whole row, we can keep the *os[1]* to be zero. Note that, the *os[0]* can be as large as 300 as described in the later discussions and FasTensor efficiently handles such large overlap.

Listing 4.1 is a complete C++ program, and users can compile it following the instruction given in Sect. 3.1. The sample code has divided the input data array into 4 blocks, therefore, it could use up to 4 MPI processes at once. The output data is stored in an HDF5 file named "das_similarity.h5".

Self-similarity calculation to identify an earthquake. To illustrate how to use the self-similarity function to identify signals from DAS data, we run the code in Listing 4.1 on a sample of DAS data. This sample comes from a DAS deployment in West Sacramento, California, USA, consisting of 25 kilometer (km) fiber-optic cable [90]. The cable runs between two cities, Sacramento and Woodland (as shown in Fig. 4.2). The 25 kilometer optic fiber can be regarded as 11,648 sensors. The sample we use covers a 6-minute interval.

During the 6-minute interval from Jan 4th 2018, it records a M4.4 earthquake south east of Berkeley.[1] In addition to the earthquake, the sample data also contains vibrations from vehicles driving nearby, water pumps, as well other source including wind and water. Since these signals affect nearby channels differently, the self-similarity calculation from *udf_similarity* could differentiate many signals. In Fig. 4.2c–e, we see that as the channel offset (*ch_ofst*) increases the earthquake signal becomes more and more distinctive because it comes from far away and affects many channels at the same time. Weaker signals from vehicles and pumping stations affect a relatively small number of channels and only visible with small (*ch_ofst*) values.

---

[1] https://earthquake.usgs.gov/earthquakes/eventpage/nc72948801/executive

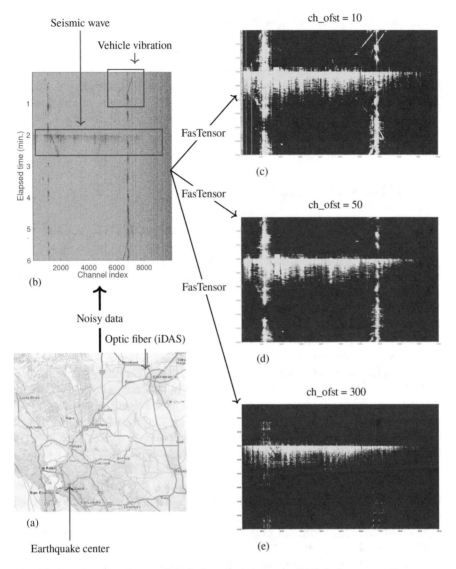

Fig. 4.2 Sample output from self-similarity calculation on a DAS deployment at Sacramento deployment in USA. (**a**) the map from the USGS website (https://earthquake.usgs.gov/earthquakes/eventpage/nc72948801/executive). The map shows the center of a magnitude 4.4 earthquake on January 4th, 2018.The iDAS deployment is in the Sacramento, north east of the Berkeley. (**b**) The raw record data from the iDAS deployment showing 10,000 channels for 6 minutes. The data is highly noisy. (**c**) Results from self-similarity calculations with *ch_ofst=10*. (**d**) Results from self-similarity calculations with *ch_ofst=50*. The vehicles' vibrations are filtered out since these vibrations are only detected by channels close together. (**e**) Results from self-similarity calculations with *ch_ofst=300*. The only visible signal is from the earthquake because it reached all channels at the same time. The *hwdth* is 500 and the *w_ofst* is 50 in (**c**), (**d**) and (**e**). The vertical stripes are two persistent noises

Other DAS data analyses in FasTensor. Reference [97] presents additional DAS data analysis functions implemented with FasTensor (under its previous name ArrayUDF). The analysis operations include decimation, Fast Fourier Transform (FFT), inverse FFT, resampling, and so on. These operations demonstrate a variety of ways to use FasTensor in large-scale data analyses using thousands of compute nodes on many terabytes of data at the same time. Such a scale was impossible with the Matlab scripts previously used by the geoscientists, FasTensor offered a new capability for analyzing DAS data.

4.2 VPIC: Vector Particle-In-Cell

VPIC and its significant tasks in scientific explorations. The plasma[2] state is the fourth state of the matter after the common states of solid, liquid, and gas. Around us, the most visible phenomena involving plasma are lightning and the aurora borealis. Plasma is a ionized state of matter, consisting of ions and electrons, first found by the chemist Irving Langmuir in the 1920s. The study of plasma has wide applications in understanding properties of heat, light, and sound that are important to formation of stars and energy production through fusion. In laboratories, scientists often produce plasma by running some neutral gas through strong electromagnetic field; and in theoretical studies of plasma, a common tool is to simulate the motion of plasma particles under similar electromagnetic field.

VPIC[3] is a 3D electromagnetic kinetic particle-in-cell plasma code [98]. We use it as a representative scientific simulation on high performance computing (HPC) systems. This simulations are critical for exploring the scientific phenomena that could not be measured directly or understood with other approximate methods. In plasma physics, one such example is magnetic reconnection and VPIC is a popular tool for studying magnetic reconnection.

VPIC splits the simulation space (or called volume) into lots of small cells, as shown in Fig. 4.3, and calculates motion of each plasma particle under the influence of electromagnetic fields defined on the mesh. The calculations not take into account of external magnetic field, but also the electromagnetic fields generated by the motion of the plasma particles. VPIC simulate the state of each particle in discrete time steps. As it determines the particle positions and velocities, it also updates the electromagnetic fields all mesh points.These calculations are based on a second-order leapfrog algorithm.

Data analysis challenges involving VPIC. In a typical run of VPIC, it may employ a mesh with millions of cells and billions of particles. Such a large VPIC simulation requires cutting-edge HPC systems with millions of CPU cores and terabyte of storage space for their data. Extracting meaningful information from terabytes of VPIC simulation data is a challenging problem.

---

[2]https://en.wikipedia.org/wiki/Plasma_(physics)

[3]https://github.com/lanl/vpic

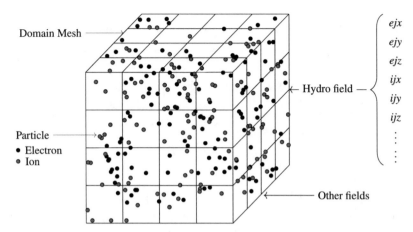

Fig. 4.3 Illustration of a 3D VPIC simulation of plasma particles, electrons and ions, under an electromagnetic field defined on a uniform mesh for electromagnetic field. In addition to the external magnetic field, the mesh is also used to record the magnetohydrodynamic field generated by the motion of plasma particles, such as the electric current and the magnetic field induced by the current. In this section, our example calculations will involve the current from electrons (*ejx*, *ejy* and *ejz*) and from ions (*ijx*, *ijy* and *ijz*)

To understand the behavior of plasma particles, a VPIC run often output diagnostics data periodically. The diagnostic data contains different data structures, for example, 1D arrays for particle attributes and 2D or 3D arrays for various fields. Many different analysis examples can be found in the published literature [99]. In the remaining of this section, we present a case study to show how to use FasTensor to extract the absolute value of current (absJ) from a magnetohydrodynamic field (and will refer to magnetohydrodynamic simply as hydro).

Calculating absolute value of current. A bare bone function for calculating absolute value of current, absJ, in FasTensor is shown in Listing 4.2. Note that VPIC computes hydro fields for electrons and ions separately. These two types of fields are stored into two different files. Each hydro field output has several attributes, named *jx*, *jy* and *jz*, for different dimensions of a 3D space. The calculation of absJ is given in user-defined function *udf_absJ* and the components of hydro fields are assembled into an array of `struct` through the function AppendAttribute presented in Sect. 3.3.6.

In function *udf_absJ*, the FasTensor **Array** object for representing the hydro fields needed for computing the current is assembled from line 28 to line 34, where the six components of two hydro fields are added one at a time using AppendAttribute. Note that these two hydro fields are stored in two files, "hydro_ele.h5" for electrons and "hydro_ion.h5" for ions. The three components of these hydro fields are recorded as *jx*, *jy*, and *jz*. From line 35 to line 39, the FasTensor **Array** object B is created to store the results. The object B has four attributes to store three components of the combined current, *jx*, *jy*, *jz* and *absJ* and its magnitude absJ. After setting the

Listing 4.2 The FasTensor implement for the absJ data analysis for VPIC

```
1     #include "ft.h"
2     struct Hydro
3     { // for input hydro fields
4         AU_UDT_INIT(Hydro)
5         float ejx, ejy, ejz;
6         float ijx, ijy, ijz;
7     };
8     struct Diag
9     { // for output current
10        AU_UDT_INIT(Diag)
11        float jx, jy, jz;
12        float absJ;
13    };
14    inline Stencil<Diag> udf_absJ(const Stencil<Hydro> &iStencil)
15    { // calculate current from input hydro fields
16        Diag dia;
17        //Total Current and Absolute Value of Current
18        dia.jx = iStencil(0, 0, 0).ejx + iStencil(0, 0, 0).ijx;
19        dia.jy = iStencil(0, 0, 0).ejy + iStencil(0, 0, 0).ijy;
20        dia.jz = iStencil(0, 0, 0).ejz + iStencil(0, 0, 0).ijz;
21        dia.absJ = sqrtf(dia.jx * dia.jx + dia.jy * dia.jy +
                  ↪dia.jz * dia.jz);
22        return Stencil<Diag>(dia);
23    }
24    int main(int argc, char *argv[])
25    { // connect to input data, assemble the data structures
             ↪needed for current calculation
26        AU_Init(argc, argv);
27        std::vector<int> cs = {4, 4, 4};
28        FT::Array<Hydro> *A = new FT::Array<Hydro>(cs);
29        A->AppendAttribute<float>("EP_HDF5:hydro_ele.h5:/jx");
30        A->AppendAttribute<float>("EP_HDF5:hydro_ele.h5:/jy");
31        A->AppendAttribute<float>("EP_HDF5:hydro_ele.h5:/jz");
32        A->AppendAttribute<float>("EP_HDF5:hydro_ion.h5:/jx");
33        A->AppendAttribute<float>("EP_HDF5:hydro_ion.h5:/jy");
34        A->AppendAttribute<float>("EP_HDF5:hydro_ion.h5:/jz");
35        FT::Array<Diag> *B = new FT::Array<Diag>();
36        B->AppendAttribute<float>("EP_HDF5:vpic_dia.h5:/jx");
37        B->AppendAttribute<float>("EP_HDF5:vpic_dia.h5:/jy");
38        B->AppendAttribute<float>("EP_HDF5:vpic_dia.h5:/jz");
39        B->AppendAttribute<float>("EP_HDF5:vpic_dia.h5:/absJ");
40        A->Transform(udf_absJ, B);
41        delete A;
42        delete B;
43        AU_Finalize();
44        return 0;
45    }
```

input and output array objects, FasTensor executes the function *udf_absJ*, which include the operations of reading from the two input data files, distributed the data, compute the output variables, and write the results to the file named *./vpic_dia.h5*.

The user-specified operations are given in function *udf_absJ*. Its input Stencil object is customized by the template argument *struct* Hydro defined in line 2. This *struct* contains member variables named *ejx*, *ejy*, *ejz*, *ijx*, *ijy*, and *ijz*, where *ejx*, *ejy* and *ejz* representing the three components of current due to electrons, and *ijx*, *ijy* and *ijz* represent the three components of current due to ions. The *udf_absJ* has the Stencil output object, whose element type is Diag. The *struct* Diag contain four elements, *jx*, *jy*, *jz* and *absJ*. Based on the *struct* Hydro and the *struct* Diag, the three components of the combined current and the magnitude of the current absJ are computed with Eqs. 4.2, 4.3, 4.4 and 4.5.

$$jx = ejx + ijx \tag{4.2}$$

$$jy = ejy + ijy \tag{4.3}$$

$$jz = ejz + ijz \tag{4.4}$$

$$absJ = \sqrt{jx^2 + jy^2 + jz^2} \tag{4.5}$$

Visualization of the sample calculation. An illustration of the procedure for computing absolute value of combined current is shown in Fig. 4.4. Even through this illustration uses a $24 \times 24 \times 24$ 3D mesh, the example code in Listing 4.2 does not restrict the mesh size. In our tests, we find this example code scales very well on HPC systems.

In Fig. 4.4, the first row contains three images illustrating three components of the current due to electrons, named *ejx*, *ejy*, and *ejz*. The second row similarly contains three components of the ion current, *ijx*, *ijy*, and *ijz*. The function *udf_absJ* computes the combined current by adding these from electrons and ions, one component at a time, and then compute the absolute values as the root-square of the three components.

Other VPIC data analyses in FasTensor. Reference [99] presents a range of analysis operations on VPIC data sets, and it also gives pseudo-codes using FasTensor to implement these data analysis operations. Listing 4.2 above performs a simple calculation on the hydro fields to produce another hydro field, the combined current. Following this calculation, this new field value could enter into another user-defined function to compute additional quantities, e.g., the variation of the current over space. Calculations on the particle data can also be easily programmed in FasTensor. Additionally, it is also possible to combine information from both particles and fields, for example, to determine the magnetic field strength at the position of each particle, which is an interpolation operation based on the magnetic field values near the particle position. This interpolation operation can be quite

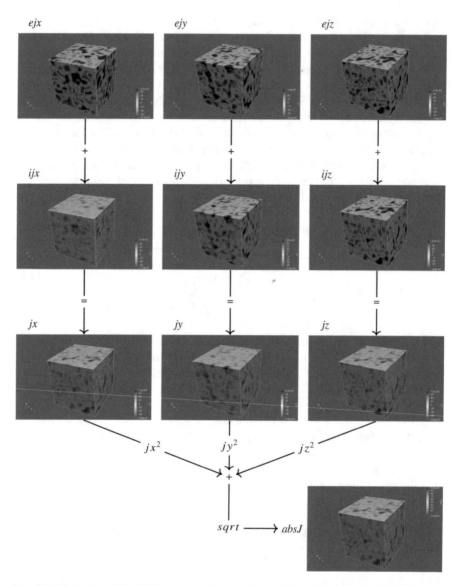

Fig. 4.4 Illustration of the VPIC data analysis example on a 24 × 24 × 24 hydro field. From top down, the first row has three components of electron current and the second row shows the iron current. The sample code in Listing 4.2 combines the two currents and then computes the absolute value and stores is in a new hydro field known as absJ

complex as the field data could be recorded on mesh points, edges or faces. Given this complexity, it is critical to efficiently locate the nearest field values. The published work also presents how to implement the global position search algorithm in FasTensor [99].

Appendix

A.1 Installation Guide

In this appendix section, we present how to install FasTensor. Listing A.1 gives an example of how to install FasTensor on the supercomputer Cori at NERSC.[1] Cori is a Cray XC40 system, with a typical linux environment similar to many supercomputers. Each node in the Cori supercomputer has a SUSE Linux Enterprise Server 15. Intel compilers are the default compilers on Cori since the system comes with Intel CPU cores. GNU GCC compilers are also available on Cori and can be used to compile FasTensor too. Listing A.1 uses the Intel C++ compiler. These installation instructions for Cori could be portable to other supercomputer systems.

Broadly, Listing A.1 first retrieves source, then configures the package, and finally creates library. More specifically, line 1 of the listing contains the command to obtain (or called "clone") the source code from the official repository of FasTensor using **git**.[2] The source code of FasTensor will be stored at the current directory under the name "fastensor". Line 3 of Listing A.1 shows the command **module**[3] to change the Intel compilers from version 19.0.3.199 to the version 19.1.3.304 with the latest C++ 2017 features. Recall that the development of the FasTensor utilizes the C++ 2017 features to support its member functions. The information about C++ 2017 feature support can be found in the table online.[4] In line 4, we again use the **module** command to load the HDF5 library named "cray-hdf5-parallel". The version of the current HDF5 is 1.10.5.2. The location of

---

[1]Cori supercomputer at NERSC: https://www.nersc.gov/systems/cori/.

[2]The **git** command comes from the Git project, a free and open source distributed version control system. More information can be found at https://git-scm.com/.

[3]The **module** command comes from the Modules package, a tool to manage shell initialization and modify the user environment using modulefiles. More information about the command **module** can be found at https://modules.readthedocs.io/en/latest/.

[4]https://en.cppreference.com/w/cpp/compiler_support

© The Author(s), under exclusive license to Springer Nature Switzerland AG 2021   85
B. Dong et al., *User-Defined Tensor Data Analysis*, SpringerBriefs in Computer
Science, https://doi.org/10.1007/978-3-030-70750-7

Listing A.1 Compile and install FasTensor on Cori supercomputer at NERSC with Intel compiler and HDF5 library

```
1  $ git clone https://bitbucket.org/berkeleylab/fastensor.git
2  $ cd fastensor
3  $ module swap  intel/19.0.3.199 intel/19.1.3.304
4  $ module load cray-hdf5-parallel
5  $ ./autogen.sh
6  $ ./configure --prefix=$PWD/build --with-hdf5=$HDF5_ROOT CXX=CC
7  $ make
8  $ make install
```

HDF5 library is encoded in the environment variable HDF5_ROOT and passed to FasTensor as shown in line 6.

In line 5, it rebuilds the compiling script via running the shell script "autogen.sh". The "autogen.sh" comes from FasTensor. It is not mandatory to run "autogen.sh" but we recommend to run it in order to create correct compiling script for specific system. In line 6, the "configure" script is executed to create the final compiling Makefile. The "configure" takes a few parameters in this case. The "CXX=CC" specifies the C++ compiler that will be used to compile the FasTensor. The "CC" actually is a wrapper of the Intel C++ compiler at the Cori supercomputer. The "–prefix" argument specifies the location to install the code. In this case, the "$PWD/build" specifies the "build" directory under the current directory. The "–with-hdf5" argument enables the HDF5 in FasTensor and also specifies the place for HDF5 library directory. Finally, at the line 7 and the line 8, the FasTensor code is compiled and installed with the "make" command. It is worth noting that the "CC" command on the Cori actually also contain the link to the MPI library. Hence, there is no need to set up the MPI for FasTensor.

In the Listing A.2, we also give an example how to install FasTensor on a Mac OS system. These installation steps are portable to other systems like Linux systems. The Listing A.2 presents steps to install the minimum requirements for FasTensor. These requirements includes MPI and HDF5.[5] The MPI implementation is the MPICH[6] but the FasTensor can also work with other implementations, such as OpenMPI.[7]

We concludes all these specific configuration options for the FasTensor in the Table A.1. As stated above, the "-with-*" option enables the corresponding component to include for FasTensor. The "–with-hdf5" is the only mandatory one and others are optional. It also has two experimental option "–enable-python" and "–with-swig=DIR" to enable the Python API for the FasTensor. The "–enable-openmp" enables the multi-threaded running of the *Transform*.

---

[5]HDF5: https://www.hdfgroup.org/

[6]MPICH: https://www.mpich.org/

[7]OpenMPI: https://www.open-mpi.org/

Listing A.2 Compile and install FasTensor and its dependents , including MPICH and HDF5, on Mac OS (version 10.14.6) from scratch. It also can serve an example to compile FasTensor on Linux systems. The line startings with the # symbol is the comment. The MPICH_ROOT, HDF5_ROOT and FT_ROOT are exported environment availables which recrod the places for the tnstallation of FasTensor, MPI and HDF5, respectively. It is worth noting that the MPICH is one of implementations of the MPI and its other implementations also work. The installation uses the GNU bash which contains the **export** command to set and manage environment availables

```
 1  ############################################################
 2  # Install MPICH at MPICH_ROOT/build.                       #
 3  # Please download it from https://www.mpich.org/downloads/ #
 4  # This example uses mpich-3.3.2.tar.gz                      #
 5  ############################################################
 6  $ tar zxvf mpich-3.3.2.tar.gz
 7  $ export MPICH_ROOT=$PWD/build
 8  $ ./configure --prefix=$MPICH_ROOT
 9  $ make
10  $ make install
11  ############################################################
12  # Install HDF5 at HDF5_ROOT/build                          #
13  # Please get the source code from                          #
14  #    https://www.hdfgroup.org/downloads/hdf5/source-code/  #
15  # The examples uses hdf5-1.12.0.tar.gz                     #
16  ############################################################
17  $ tar zxvf hdf5-1.12.0.tar.gz
18  $ cd hdf5-1.12.0
19  $ export HDF5_ROOT=$PWD/build
20  $ ./configure --enable-parallel --prefix=$HDF5_ROOT
        ↪CC=$MPICH_ROOT/bin/mpicc
21  $ make
22  $ make install
23  #####################################
24  # Install FasTensor at FT_ROOT/build #
25  #####################################
26  $ git clone https://bitbucket.org/berkeleylab/fastensor.git
27  $ cd fastensor
28  $ export FT_ROOT=$PWD/build/
29  $ ./autogen.sh
30  $ ./configure --prefix=$FT_ROOT --with-hdf5=$HDF5_ROOT
        ↪CXX=$MPICH_ROOT/bin/mpicxx
31  $ make
32  $ make install
```

Table A.1 Summary of the specific configurations for FasTensor during compiling. * denotes these function are experimental ones now

Configuration option	Function
–enable-debug	enable debug (-g3)
–enable-openmp	enable OpenMP on the Transform function
–with-hdf5=DIR	use DIR as the root directory for HDF5 include and lib
–with-adios=DIR	use DIR as the root directory for ADIOS include and lib
–with-pnetcdf=DIR	use DIR as the root directory for PnetCDF include and lib
–with-dash=DIR	use DIR as the root directory for DASH include and lib
–prefix=DIR	user DIR as the root directory to install FasTensor.
–enable-python*	enable python API for FasTensor
–with-swig=DIR*	use DIR as the root directory for swig's include and lib

A.2   How to Develop a New Endpoint Protocol

The Endpoint in FasTensor represents the data sources as well as the data analysis outputs. By default, the FasTensor supports the Endpoint protocol for HDF5, ADIOS, PNetCDF, Memory(DASH), Binary, and TDMS data file. Although these data file formats cover most scientific data files, other file formats may exist and are used by the scientific community. In this section, we give a high-level guideline about how to add a new endpoint protocol for a new file format. The FasTensor utilizes the factory design pattern to support different endpoint protocols. The Listing A.3 gives a summary of a few key methods and members for the Endpoint class. The Endpoint class is the base class in the factory design pattern. A new endpoint protocol can be declared as a new C++ class which takes the Endpoint class as the base class.

In other words, the new endpoint protocol shares the same class members as the Endpoint class but it can have different implementation of these methods. All these class members are called or used by FasTensor to run its **Transform** function since it is important to have an implementation in the new endpoint protocol. To enforce this purpose, the Endpoint class declares most of its methods as pure virtual methods with the **virtual** key word at the front and the **= 0** symbol at the end. These pure virtual methods require to be overwritten in an derived new endpoint protocol. The constructor function *Endpoint( )* is required to be the normal function by the C++ standard. There are few functions, including *WriteAttribute*, *ReadAttribute* and *Control*, are just regular virtual functions. These regular functions can be overlooked by the new endpoint.

The Endpoint class has at lest three attributes **endpoint_info**, **endpoint_dim_size** and **data_element_type**. The **endpoint_info** actually contains the information extracted from the **endpoint_id** from the calling Array class. The **endpoint_info** is set by the FasTensor when it creates the Endpoint object. Basically, the FasTensor separates the type from the **endpoint_id**. Taking the "EP_HDF5:./tutorial.h5:/dat" as example **endpoint_id**, the FasTensor splits it into two parts, i.e., "EP_HDF5"

Listing A.3 List of major virtual functions for an the Endpoint class for users to develop their own endpoint protocol. Most of its methods are denoted as pure virtual functions (with the *=0* sign) that must be implemented in the new endpoint protocol because these methods are called by the FasTensor. One can keep their function body empty if these methods are not required by the new endpoint

```
1    class Endpoint
2    {
3     protected:
4       string endpoint_info; //set during initialization
5       vector<unsigned long long> endpoint_size;
6       FTType data_element_type;
7     public:
8       Endpoint(){};
9       virtual ~Endpoint(){};
10      //pure virtual methods
11      virtual int ParseEndpointInfo() = 0;
12      virtual int Map2MyType() = 0;
13      virtual int ExtractMeta() = 0;
14      virtual int Create() = 0;
15      virtual int Open() = 0;
16      virtual int Read(vector<unsigned long long> start,
             ↪vector<unsigned long long> end, void *data_p) = 0;
17      virtual int Write(vector<unsigned long long> start,
             ↪vector<unsigned long long> end, void *data_p) = 0;
18      virtual int Close() = 0;
19      //virtual methods
20      virtual int WriteAttribute(string name, void *data_p,
             ↪FTType type, size_t length);
21      virtual int ReadAttribute(string  name, void *data_p,
             ↪FTType type, size_t length);
22      virtual int Control(int cmd, vector<string> &argv));
23   };
```

and "./tutorial.h5:/dat". Then, the FasTensor assigns the "./tutorial.h5:/dat" to the *endpoint_info* in the Endpoint class. The *endpoint_dim_size* contains the size of the endpoint. The *endpoint_dim_size* is a vector with size *n* that contains the size of the data in the endpoint. The *endpoint_dim_size* can be set by the *ExtractMeta*, which needs to be implemented in the new endpoint protocol. The *data_element_type* contains the type for the data element which is set by the FasTensor after creating the Endpoint object.

The Endpoint class has a constructor *Endpoint()* to perform some initialization work. The implementation for the new Endpoint protocol does not need to implement it but it can has its own constructor. The destructor of the FasTensor is a virtual function. This virtual destructor guarantees that the object of derived new endpoint protocol is destroyed properly. In this case, the destructors from the base Endpoint class and the derived new Endpoint class can be called during the free of the object.

The *ParseEndpointInfo* converts the **endpoint_info** into the endpoint-specific information. It parses the **endpoint_info** string to extract essential information from it for other methods to use. The FasTensor sets the **endpoint_info** string when it creates the object. The should be called by the Constructor of the new Endpoint object. For example, in the HDF5 endpoint class, the *ParseEndpointInfo* can split the "./tutorial.h5:/dat" into two parts, "./tutorial.h5" and "/dat". The former is the HDF5 file name and the later is the HDF5 data set name and its group path.

The *Map2MyType* converts the **data_element_type** in FasTensor to the endpoint specific one. The data type in FasTensor are concluded in the Table 3.3 in the previous section. The table also presents its relationships to C++ and other endpoint's data element types. The **data_element_type** is set by the FasTensor after creating the object. The *Map2MyType* should be called before the type is used within the other methods, such as *Read* and Write.

The *ExtractMeta* extracts the metadata information for the target endpoint (mostly as input). One essential parameter needs to be set by the *ExtractMeta* is the **endpoint_dim_size**. Given the specific file type, the *ExtractMeta* may call the *Open* method to open the target object first and then find the metadata. Other metadata can also be added by the new endpoint protocol. For example, in the TDMS endpoint in the FasTensor, it has its own metadata to record the offset of the actual data after the metadata header.

The *Create* creates the target endpoint if necessary. Usually, this function is called to store the output data from the *Transform*. For example, the *Create* method in the HDF5 Endpoint creates a HDF5 file for the *Transform* output. The *Open* method opens the target endpoint for the following *Read* and Write methods. In certain endpoint, such as the Binary endpoint, the *Create* and the *Open* can be merged since it has a single API for both the *Create* and the *Open* semantics. After the endpoint is created or opened, the *SetOpenFlag* can be called to set the open flag on the endpoint. The *Close* closes the endpoint and also release the resources. Certain functionality for resource release can also be put into the destructor method.

The *Read* and the Write for the Endpoint class have the following signatures. Both the *Read* and the Write methods take there parameter, **start**, **end** and **data_p**. Both **start** and the **end** are vector with the size n. The **start** and the **end** contains the address to read or write the data. The **data_p** is an address pointer for the actual data. During the *Read* function, the **data_p** contains the address pointer for the data that are read from the endpoint. During the *Write* function, the **data_p** contains the address pointer for the data that are written to the endpoint. The actual data elements can be ordered by the new Endpoint to match the actual target file format.

```
virtual int Read(vector<unsigned long long> start,
    ↪vector<unsigned long long> end, void *data_p) = 0;
virtual int Write(vector<unsigned long long> start,
    ↪vector<unsigned long long> end, void *data_p) = 0;
```

The *WriteAttribute* and the *ReadAttribute* have the following API signatures. The *WriteAttribute* writes a key (i.e., **name**))indexed attribute to the endpoint. The *ReadAttribute* reads a key (i.e., **name**))indexed attribute from the endpoint. The *data_p* contains the address pointer for the data either to read or write. Being different from classic key-value pair, the *data_p* can have as many as **length** data which has the *type*. Both *WriteAttribute* and *ReadAttribute* are regular virtual function and new endpoint can choose to have an implementation for it or not.

```
virtual int WriteAttribute(string name, void *data_p, FTDataType
    ↪type, size_t length);
virtual int ReadAttribute(string  name, void *data_p, FTDataType
    ↪type, size_t length);
```

The *Control* answers the *ControlEndpoint* call from the **Array** class object (as it presented in the previous Sect. 3.3.8). Being the same as the *ControlEndpoint*, the *Control* takes two parameters, *cmd* and *argv*, as shown by the below code signature. The *cmd* is the command (integer type) which is defined specifically by the new endpoint. The *argv* is a vector of string which contains arguments for the *cmd* defined by an endpoint. The *argv* can be kept empty if there is no arguments needed for the *cmd*. The *argv* also contain the output value if the *cmd* has output.

```
virtual int Control(int cmd, vector<string> &argv);
```

Bibliography

1. T. Hey, S. Tansley, K. Tolle (eds.), *The Fourth Paradigm: Data-Intensive Scientific Discovery* (Microsoft, Redmond, 2009)
2. ALEPH Collaboration, DELPHI Collaboration, L3 Collaboration, OPAL Collaboration, and The LEP Working Group for Higgs Boson Searches, Search for the standard model higgs boson at LEP. Phys. Lett. B **565**, 61–75 (2003). [Online]. Available: http://www.sciencedirect.com/science/article/pii/S0370269303006142
3. B.P. Abbott, R. Abbott, T.D. Abbott, M.R. Abernathy, F. Acernese, K. Ackley, C. Adams, T. Adams, P. Addesso, R.X. Adhikari, V.B. Adya, C. Affeldt, M. Agathos, K. Agatsuma, N. Aggarwal, O.D. Aguiar, L. Aiello, A. Ain, P. Ajith, B. Allen, A. Allocca, P.A. Altin, S.B. Anderson, W.G. Anderson, K. Arai, M.A. Arain, M.C. Araya, C.C. Arceneaux, J.S. Areeda, N. Arnaud, K.G. Arun, S. Ascenzi, G. Ashton, M. Ast, S.M. Aston, P. Astone, P. Aufmuth, C. Aulbert, S. Babak, P. Bacon, M.K.M. Bader, P.T. Baker, F. Baldaccini, G. Ballardin, S.W. Ballmer, J.C. Barayoga, S.E. Barclay, B.C. Barish, D. Barker, F. Barone, B. Barr, L. Barsotti, M. Barsuglia, D. Barta et al., Observation of gravitational waves from a binary black hole merger. Phys. Rev. Lett. **116**, 061102 (2016). [Online]. Available: https://doi.org/10.1103/PhysRevLett.116.061102
4. M.M. Kasliwal, E. Nakar, L.P. Singer, D.L. Kaplan, D.O. Cook, A. Van Sistine, R.M. Lau, C. Fremling, O. Gottlieb, J.E. Jencson, S.M. Adams, U. Feindt, K. Hotokezaka, S. Ghosh, D.A. Perley, P.-C. Yu, T. Piran, J.R. Allison, G.C. Anupama, A. Balasubramanian, K.W. Bannister, J. Bally, J. Barnes, S. Barway, E. Bellm, V. Bhalerao, D. Bhattacharya, N. Blagorodnova, J.S. Bloom, P.R. Brady, C. Cannella, D. Chatterjee, S.B. Cenko, B.E. Cobb, C. Copperwheat, A. Corsi, K. De, D. Dobie, S.W.K. Emery, P.A. Evans, O.D. Fox, D.A. Frail, C. Frohmaier, A. Goobar, G. Hallinan, F. Harrison, G. Helou, T. Hinderer, A.Y.Q. Ho, A. Horesh, W.-H. Ip, R. Itoh, D. Kasen, H. Kim, N.P.M. Kuin, T. Kupfer, C. Lynch, K. Madsen, P.A. Mazzali, A.A. Miller, K. Mooley, T. Murphy, C.-C. Ngeow, D. Nichols, S. Nissanke, P. Nugent, E.O. Ofek, H. Qi, R.M. Quimby, S. Rosswog, F. Rusu, E.M. Sadler, P. Schmidt, J. Sollerman, I. Steele, A.R. Williamson, Y. Xu, L. Yan, Y. Yatsu, C. Zhang, W. Zhao, Illuminating gravitational waves: a concordant picture of photons from a neutron star merger. Science (2017). [Online]. Available: https://science.sciencemag.org/content/early/2017/10/13/science.aap9455
5. B. Dong, P. Kilian, X. Li, F. Guo, S. Byna, K. Wu, Terabyte-scale particle data analysis: an arrayudf case study, in *Proceedings of the 31st International Conference on Scientific and Statistical Database Management, SSDBM 2019*, Santa Cruz, 23–25 July 2019, pp. 202–205. [Online]. Available: https://doi.org/10.1145/3335783.3335805

6. B. Dong, V.R. Tribaldos, X. Xing, S. Byna, J. Ajo-Franklin, K. Wu, DASSA: parallel DAS data storage and analysis for subsurface event detection, in *2020 IEEE International Parallel and Distributed Processing Symposium (IPDPS)* (IEEE, 2020), pp. 254–263

7. B. Dong, K. Wu, S. Byna, H. Tang, Slope: structural locality-aware programming model for composing array data analysis, in *High Performance Computing*, ed. by M. Weiland, G. Juckeland, C. Trinitis, P. Sadayappan (Springer International Publishing, Cham, 2019), pp. 61–80

8. B. Dong, K. Wu, S. Byna, J. Liu, W. Zhao, F. Rusu, Arrayudf: user-defined scientific data analysis on arrays, in *Proceedings of the 26th International Symposium on High-Performance Parallel and Distributed Computing*. HPDC'17 (ACM, New York, 2017), pp. 53–64. [Online]. Available: https://doi.org/10.1145/3078597.3078599

9. C. Alcock et al., The MACHO project: microlensing results from 5.7 years of LMC observations. Astrophys. J. **542**, 281–307 (2000)

10. C.J. Grillmair, R. Laher, J. Surace, S. Mattingly, E. Hacopians, E. Jackson, J. van Eyken, B. McCollum, S. Groom, W. Mi, H. Teplitz, An overview of the palomar transient factory pipeline and archive at the infrared processing and analysis center, in *Astronomical Data Analysis Software and Systems XIX*, ed. by Y. Mizumoto, K.I. Morita, M. Ohishi. Astronomical Society of the Pacific Conference Series, vol. 434. (Astronomical Society of the Pacific, San Francisco, 2010), p. 28

11. R.I. Cutri et al., Explanatory supplement to the 2mass all sky data release and extended mission products, https://old.ipac.caltech.edu/2mass/releases/allsky/doc/ (2008)

12. V. Kumar, R.L. Grossman, *Data Mining for Scientific and Engineering Applications* (Kluwer Academic Publishers, Dordrecht/Boston, 2001)

13. R. White, A. Rest et al., The panstarrs dr1 data release, https://archive.stsci.edu/mug/mug_2016-Dec/20MUG_PS1_White_2016dec.pdf (2016)

14. I. Foster, Transformative role of computation and 'big data', https://www.energy.gov/sites/prod/files/SEAB%20-%20Foster%20presentation.pdf (2012)

15. J. Dean, S. Ghemawat, MapReduce: simplified data processing on large clusters, in *OSDI'04* (2004)

16. J. Dean, S. Ghemawat, MapReduce: simplified data processing on large clusters. Commun. ACM **51**(1), 107–113 (2008). [Online]. Available: https://doi.org/10.1145/1327452.1327492

17. T. White, *Hadoop – The Definitive Guide: MapReduce for the Cloud* (O'Reilly, 2009). [Online]. Available: http://www.oreilly.de/catalog/9780596521974/index.html

18. M. Zaharia, M. Chowdhury, T. Das, A. Dave, J. Ma, M. McCauley, M.J. Franklin, S. Shenker, I. Stoica, Resilient distributed datasets: a fault-tolerant abstraction for in-memory cluster computing, in *NSDI 2012* (2012)

19. M. Zaharia, R.S. Xin, P. Wendell, T. Das, M. Armbrust, A. Dave, X. Meng, J. Rosen, S. Venkataraman, M.J. Franklin, A. Ghodsi, J. Gonzalez, S. Shenker, I. Stoica, Apache spark: a unified engine for big data processing. Commun. ACM **59**(11), 56–65 (2016). [Online]. Available: https://doi.org/10.1145/2934664

20. M. Abadi, P. Barham, J. Chen, Z. Chen, A. Davis, J. Dean, M. Devin, S. Ghemawat, G. Irving, M. Isard, M. Kudlur, J. Levenberg, R. Monga, S. Moore, D.G. Murray, B. Steiner, P. Tucker, V. Vasudevan, P. Warden, M. Wicke, Y. Yu, X. Zheng, Tensorflow: a system for large-scale machine learning, in *12th USENIX Symposium on Operating Systems Design and Implementation (OSDI 16)* (USENIX Association, Savannah, 2016), pp. 265–283. [Online]. Available: https://www.usenix.org/conference/osdi16/technical-sessions/presentation/abadi

21. P. O'Neil, E. O'Neil, *Database—Principles, Programming and Performance*, 2nd edn. (Morgan Kaufmann Publishers Inc., San Francisco, CA, USA, 2000)

22. M.T. Ozsu, *Principles of Distributed Database Systems*, 3rd edn. (Prentice Hall Press, Upper Saddle River, 2007)

23. R. Brun, F. Rademakers, Root: an object oriented data analysis framework. Nucl. Instrum. Methods Phys. Res. Sect. A **289**(1–2), 81–86 (1997)

24. M. Ballintijn, R. Brun, F. Rademakers, G. Roland, The PROOF distributed parallel analysis framework based on ROOT, in *Proceedings of CHEP03*, 2003, also available from ArXiv at http://arxiv.org/abs/physics/0306110.

25. W. Gropp, E. Lusk, A. Skjellum, *Using MPI: Portable Parallel Programming with the Message–Passing Interface*, 2nd edn. (The MIT Press, Cambridge, MA, 1999)

26. M. Snir, S.W. Otto, S. Huss-Lederman, D.W. Walker, J. Dongarra, *MPI: The Complete Reference. Volume 1, The MPI-1 Core*, 2nd edn (MIT Press, Cambridge, MA, 1998). [Online]. Available: http://mitpress.mit.edu/book-home.tcl?isbn=0262692155

27. B.W. Kernighan, R. Pike, *The Unix Programming Environment* (Prentice-Hall, Englewood Cliffs, 1984)

28. D. Hildebrand, A. Nisar, R. Haskin, PNFS, POSIX, and MPI-IO: a tale of three semantics, in *Proceedings of the 4th Annual Workshop on Petascale Data Storage*. PDSW'09 (Association for Computing Machinery, New York, 2009), pp. 32–36. [Online]. Available: https://doi.org/10.1145/1713072.1713082

29. M. Vilayannur, S. Lang, R. Ross, R. Klundt, Ward, Extending the posix I/O interface: a parallel file system perspective. Technical Report ANL/MCS-TM-302, 12 (2008)

30. D.C. Wells, E.W. Greisen, R.H. Harten, FITS: a flexible image transport system. Astron. Astrophys. Suppl. Ser. **44**, 363–370 (1981)

31. R. Rew, G. Davis, NetCDF: an interface for scientific data access. IEEE Comput. Graph. Appl. **10**(4), 76–82 (1990), software available at http://www.unidata.ucar.edu/software/netcdf/

32. M. Folk, G. Heber, Q. Koziol, E. Pourmal, D. Robinson, An overview of the HDF5 technology suite and its applications, in *Proceedings of the EDBT/ICDT 2011 Workshop on Array Databases* (ACM, 2011), pp. 36–47, software at http://www.hdfgroup.org/HDF5/

33. J. Dean, S. Ghemawat, Mapreduce: simplified data processing on large clusters. Commun. ACM **51**(1), 107–113 (2008)

34. M. Factor, K. Meth, D. Naor, O. Rodeh, J. Satran, Object storage: the future building block for storage systems, in *2005 IEEE International Symposium on Mass Storage Systems and Technology* (IEEE, 2005), pp. 119–123

35. M. Stonebraker, J. Becla, D.J. DeWitt, K.-T. Lim, D. Maier, O. Ratzesberger, S.B. Zdonik, Requirements for science data bases and SciDB, in *CIDR*, vol. 7 (2009), pp. 173–184

36. P. Baumann, A. Dehmel, P. Furtado, R. Ritsch, N. Widmann, The multidimensional database system rasDaMan. SIGMOD Rec. **27**(2), 575–577 (1998). [Online]. Available: https://doi.org/10.1145/276305.276386

37. A. Shoshani, D. Rotem (eds.), *Scientific Data Management: Challenges, Technology, and Deployment* (Chapman & Hall/CRC Press, London, 2010)

38. W.F. Godoy, N. Podhorszki, R. Wang, C. Atkins, G. Eisenhauer, J. Gu, P. Davis, J. Choi, K. Germaschewski, K. Huck et al., ADIOS 2: the adaptable input output system. A framework for high-performance data management. SoftwareX **12**, 100561 (2020)

39. J.F. Lofstead, S. Klasky, K. Schwan, N. Podhorszki, C. Jin, Flexible IO and integration for scientific codes through the adaptable IO system (ADIOS), in *CLADE'08* ACM, New York, 2008), pp. 15–24. [Online]. Available: https://doi.org/10.1145/1383529.1383533

40. X. Li, F. Guo, H. Li, J. Birn, The roles of fluid compression and shear in electron energization during magnetic reconnection (2018). [Online]. Available: https://arxiv.org/pdf/1801.02255.pdf

41. N. Carriero, D. Gelernter, How to write parallel programs: a guide to the perplexed. ACM Comput. Surv. **21**(3), 323–357 (1989). [Online]. Available: https://doi.org/10.1145/72551.72553

42. J.T. Feo, *Comparative Study of Parallel Programming Languages: The Salishan Problems* (Elsevier Science Inc., North-Holland, Amsterdam, 1992)

43. B. Schmidt, J. Gonzalez-Dominguez, C. Hundt, M. Schlarb, *Parallel Programming: Concepts and Practice* (Morgan Kaufmann, Amsterdam, 2017)

44. K.M. Chandy, C. Kesselman, Compositional C++: compositional parallel programming, in *International Workshop on Languages and Compilers for Parallel Computing* (Springer, 1992), pp. 124–144

45. C.A.R. Hoare, Communicating sequential processes. Commun. ACM **21**(8), 666–677 (1978) [Online]. Available: https://doi.org/10.1145/359576.359585

46. J. Fang, C. Huang, T. Tang, Z. Wang, Parallel programming models for heterogeneous many-cores: A comprehensive survey. CCF Trans. HPC **2**, 382–400 (2020) https://doi.org/10.1007/s42514-020-00039-4

47. R.S. Bird, P.L. Wadler, Functional Programming (Prentice Hall, Hemel Hempstead, England, 1988)

48. S.A. Kamil, Productive high performance parallel programming with auto-tuned domain-specific embedded languages. Ph.D. dissertation, USA (2012)

49. B. Lisper, Data parallelism and functional programming, in *The Data Parallel Programming Model* (Springer, Berlin/Heidelberg, 1996), pp. 220–251

50. N. Makrynioti, V. Vassalos, Declarative data analytics: A survey. IEEE Trans. Knowl. Data Eng. **33**(6), 2392–2411 (2021). https://doi.org/10.1109/TKDE.2019.2958084

51. S. Thompson, *The Haskell: The Craft of Functional Programming*, 2nd edn. (Addison-Wesley Longman Publishing Co., Inc., USA, 1999)

52. G.-R. Perrin, A. Darte, *The Data Parallel Programming Model: Foundations, HPF Realization, and Scientific Applications* (Springer, Berlin, Heidelberg, 1996)

53. H. Kaiser, T. Heller, B. Adelstein-Lelbach, A. Serio, D. Fey, HPX: a task based programming model in a global address space, in *Proceedings of the 8th International Conference on Partitioned Global Address Space Programming Models. PGAS'14* (Association for Computing Machinery, 2014). [Online]. Available: https://doi.org/10.1145/2676870.2676883

54. A.T. Tan, J. Falcou, D. Etiemble, H. Kaiser, Automatic task-based code generation for high performance domain specific embedded language. Int. J. Parallel Program. **44**(3), 449–465 (2016)

55. P. Thoman, K. Dichev, T. Heller, R. Iakymchuk, X. Aguilar, K. Hasanov, P. Gschwandtner, P. Lemarinier, S. Markidis, H. Jordan et al., A taxonomy of task-based parallel programming technologies for high-performance computing. J. Supercomput. **74**(4), 1422–1434 (2018)

56. I. Watson, V. Woods, P. Watson, R. Banach, M. Greenberg, J. Sargeant, Flagship: a parallel architecture for declarative programming, in *[1988] The 15th Annual International Symposium on Computer Architecture. Conference Proceedings* (1988), pp. 124–130

57. Q. Chen, A. Therber, M. Hsu, H. Zeller, B. Zhang, R. Wu, Efficiently support mapreduce-like computation models inside parallel DBMS, in *IDEAS'09* (ACM, New York, 2009), pp. 43–53

58. E. Friedman, P. Pawlowski, J. Cieslewicz, SQL/mapreduce: a practical approach to self-describing, polymorphic, and parallelizable user-defined functions, in *VLDB*, vol. 2. (VLDB Endowment, 2009), pp. 1402–1413

59. L.J. Durlofsky, B. Engquist, S. Osher, Triangle based adaptive stencils for the solution of hyperbolic conservation laws. J. Comput. Phys. **98**(1), 64–73 (1992). [Online]. Available: http://www.sciencedirect.com/science/article/pii/002199919290173V

60. C. Johnson, *Numerical Solution of Partial Differential Equations by Finite Element Method* (Cambridge University Press, Cambridge/New York, 1987)

61. K.W. Morton, D.F. Mayers, *Numerical Solution of Partial Differential Equations.* (Cambridge University Press, Cambridge/New York, 1994)

62. F. Luporini, M. Louboutin, M. Lange, N. Kukreja, P. Witte, J. Hückelheim, C. Yount, P.H.J. Kelly, F.J. Herrmann, G.J. Gorman, Architecture and performance of devito, a system for automated stencil computation. ACM Trans. Math. Softw. **46**(1) (2020). [Online]. Available: https://doi.org/10.1145/3374916

63. M. Boehm, M.W. Dusenberry, D. Eriksson, A.V. Evfimievski, F.M. Manshadi, N. Pansare, B. Reinwald, F.R. Reiss, P. Sen, A.C. Surve, S. Tatikonda, Systemml: declarative machine learning on spark. Proc. VLDB Endow. **9**(13), 1425–1436 (2016). [Online]. Available: https://doi.org/10.14778/3007263.3007279

64. X. Meng, J. Bradley, B. Yavuz, E. Sparks, S. Venkataraman, D. Liu, J. Freeman, D. Tsai, M. Amde, S. Owen, D. Xin, R. Xin, M.J. Franklin, R. Zadeh, M. Zaharia, A. Talwalkar, MLlib: machine learning in apache spark. J. Mach. Learn. Res. **17**(1), 1235–1241 (2016)

65. Christopher Olah, Neural Networks, Types, and Functional Programming, (2015)

66. Y. Cheng, F. Rusu, Parallel in-situ data processing with speculative loading, in *Proceedings of the 2014 ACM SIGMOD International Conference on Management of Data* (2014), pp. 1287–1298

67. S. Blanas, K. Wu, S. Byna, B. Dong, A. Shoshani, Parallel data analysis directly on scientific file formats, in *SIGMOD'2014* (2014)

68. H. Xing, S. Floratos, S. Blanas, S. Byna, Prabhat, K. Wu, P. Brown, Arraybridge: interweaving declarative array processing with high-performance computing. CoRR, abs/1702.08327 (2017). [Online]. Available: http://arxiv.org/abs/1702.08327

69. A.P. Marathe, K. Salem, A Language for Manipulating Arrays, in *VLDB'97* (1997)

70. J.S. Bloom, J.W. Richards et al., Automating discovery and classification of transients and variable stars in the synoptic survey era. PASP **124**(921), 1175–1196 (2012)

71. S. Byna, J. Chou, O. Rübel, Prabhat, H. Karimabadi et al., Parallel I/O, analysis, and visualization of a trillion particle simulation, in *SC* (2012)

72. Y. Cheng, F. Rusu, Formal representation of the SS-DB benchmark and experimental evaluation in EXTASCID. Distrib. Parallel Databases **33**(3), 277–317 (2015)

73. P.G. Brown, Overview of SciDB: large scale array storage, processing and analysis, in *SIGMOD* (2010)

74. The HDF Group, HDF5 User Guide (2010)

75. J. Li, W. keng Liao, A. Choudhary, R. Ross, R. Thakur, W. Gropp, R. Latham, A. Siegel, B. Gallagher, M. Zingale, Parallel netCDF: a high-performance scientific I/O interface, in *SC'03* (ACM, New York, 2003), p. 39. [Online]. Available: https://doi.org/10.1145/1048935.1050189

76. L.T. Yang, M. Guo, *High-Performance Computing: Paradigm and Infrastructure* (Wiley, San Francisco, CA, 2005)

77. D. Orchard, A. Mycroft Efficient and correct stencil computation via pattern matching and static typing. Electron. Proc. Theor. Comput. Sci. **66**, 68–92 (2011). [Online]. Available: https://doi.org/10.4204/EPTCS.66.4

78. E. Soroush, M. Balazinska, D. Wang, ArrayStore: a storage manager for complex parallel array processing, in *SIGMOD'2011* (ACM, 2011). [Online]. Available: https://doi.org/10.1145/1989323.1989351

79. N. Chaimov, A. Malony, S. Canon, C. Iancu et al., Scaling Spark on HPC systems, in *HPDC 2016* (2016)

80. S. Chaudhuri, V.R. Narasayya, An efficient cost-driven index selection tool for Microsoft SQL server, in *VLDB'97* (1997)

81. P.C. Zikopoulos, R.B. Melnyk, *DB2: The Complete Reference* (McGraw-Hill, Inc., New York, 2001)

82. M. Widenius, D. Axmark, *MySQL Reference Manual* (O'Reilly & Associates, Inc., Sebastopol, 2002)

83. B. Momjian, *PostgreSQL: Introduction and Concepts* (Addison-Wesley Longman Publishing Co., Inc., Boston, 2001)

84. M. Raasveldt, Vectorized UDFs in column-stores (master thesis) (2015). [Online]. Available: http://dspace.library.uu.nl/handle/1874/325669

85. W. Gropp, E. Lusk, N. Doss, A. Skjellum, A high-performance, portable implementation of the MPI message passing interface standard. Parallel Comput. **22**(6), 789–828 (1996)

86. J. Meng, K. Skadron, Performance modeling and automatic ghost zone optimization for iterative stencil loops on gpus, in *Proceedings of the 23rd International Conference on Supercomputing*. ICS'09 (2009), pp. 256–265. [Online]. Available: https://doi.org/10.1145/1542275.1542313

87. M. Wehner, Prabhat, K.A. Reed, D. Stone, W.D. Collins, J. Bacmeister, Resolution dependence of future tropical cyclone projections of cam5.1 in the u.s. clivar hurricane working group idealized configurations. J. Climate **28**(10), 3905–3925 (2015)

88. E. Racah, C. Beckham, T. Maharaj, S. Ebrahimi Kahou, M. Prabhat, C. Pal, Extremeweather: a large-scale climate dataset for semi-supervised detection, localization, and understanding of extreme weather events, in *NIPS 2017* (2017)

89. J. Liu, E. Racah, Q. Koziol et al., H5Spark: bridging the I/O gap between Spark and scientific data formats on HPC systems, in *Cray User Group* (2016)

90. J. Ajo-Franklin, S. Dou, T. Daley, B. Freifeld, M. Robertson, C. Ulrich, T. Wood, I. Eckblaw, N. Lindsey, E. Martin et al., Time-lapse surface wave monitoring of permafrost thaw using distributed acoustic sensing and a permanent automated seismic source, in *SEG Technical Program Expanded Abstracts 2017* (Society of Exploration Geophysicists, 2017)

91. A. Hartog, *An Introduction to Distributed Optical Fibre Sensors* (CRC Press, Boca Raton, 2017)

92. S. Dou, N. Lindsey, A.M. Wagner, T.M. Daley, B. Freifeld, M. Robertson, J. Peterson, C. Ulrich, E.R. Martin, J.B. Ajo-Franklin, Distributed acoustic sensing for seismic monitoring of the near surface: a traffic-noise interferometry case study. Sci. Rep. **7**(1) (2017)

93. N.J. Lindsey, E.R. Martin, D.S. Dreger, B. Freifeld, S. Cole, S.R. James, B.L. Biondi, J.B. Ajo-Franklin, Fiber-optic network observations of earthquake wavefields. Geophys. Res. Lett. **44**(23), 11–792 (2017)

94. Reportlinker, Distributed acoustic sensing systems (DAS) market 2014–2024: Fibre optics for oil and gas, utility, military, infrastructure and security applications, Reportlinker, Technical Report (2014)

95. Z. Li, Z. Peng, D. Hollis, L. Zhu, J. McClellan, High-resolution seismic event detection using local similarity for large-n arrays. Sci. Rep. **8**(1), 1646 (2018)

96. X. Xing, B. Dong, J. Ajo-Franklin, K. Wu, Automated parallel data processing engine with application to large-scale feature extraction, in *2018 IEEE/ACM Machine Learning in HPC Environments (MLHPC)* (2018), pp. 37–46

97. B. Dong, V.R. Tribaldos, X. Xing, S. Byna, J. Ajo-Franklin, K. Wu, Dassa: parallel das data storage and analysis for subsurface event detection, in *2020 IEEE International Parallel and Distributed Processing Symposium (IPDPS)* (2020), pp. 254–263

98. K.J. Bowers, B.J. Albright, B. Bergen, L. Yin, K.J. Barker, D.J. Kerbyson, 0.374 pflop/s trillion-particle kinetic modeling of laser plasma interaction on roadrunner, in *Proceedings of the 2008 ACM/IEEE Conference on Supercomputing. SC'08* (IEEE Press, 2008)

99. B. Dong, P. Kilian, X. Li, F. Guo, S. Byna, K. Wu, Terabyte-scale particle data analysis: an arrayudf case study, in *Proceedings of the 31st International Conference on Scientific and Statistical Database Management. SSDBM'19* (Association for Computing Machinery, New York, 2019), pp. 202–205. [Online]. Available: https://doi.org/10.1145/3335783.3335805

Alphabetical Index

Printed in the United States
by Baker & Taylor Publisher Services